Altered Genes II

Altered Genes II

The future?

Edited by
Richard Hindmarsh & Geoffrey Lawrence

Scribe Publications
Melbourne

Scribe Publications Pty Ltd
PO Box 287
Carlton North, Victoria Australia 3054
Email: scribe@bigpond.net.au

First published by Scribe Publications Pty Ltd 2001

Typeset in Times New Roman by the publisher
Printed and bound in Australia
by Australian Print Group, Maryborough, Victoria

National Library of Australia
Cataloguing-in-Publication data

Altered genes II : the future?

2nd ed.
Bibliography.
Includes index.
ISBN 0 908011 59 8.

1. Genetic engineering - Environmental aspects - Australia.
2. Genetic engineering - Social aspects - Australia. 3.
Genetic engineering - Political aspects - Australia. 4.
Genetic engineering - Australia - Moral and ethical aspects.
I. Hindmarsh, R. A. II. Lawrence, G. A.

174.957

The image on the front cover is of cell mitosis

www.scribepub.com.au

Contents

Part 3 Molecular farming, biopiracy & campaigning

Tables and Figures

Contributors

Jean Christie is director of international liaison for RAFI (the Rural Advancement Foundation International), an international non-governmental organisation based in Ottawa, Canada. For over twenty years RAFI has undertaken research and policy advocacy on issues relating to biodiversity, biotechnology, intellectual property rights, and indigenous peoples' knowledge. With a BA in political science from Queens University (Canada) Jean was—prior to joining RAFI—executive director of Inter Pares, a Canadian non-profit organisation working in international cooperation for development.

Stephen Crook is Foundation Professor of Sociology at James Cook University (Queensland), and is President of the Australian Sociological Association (1998-2002). He has written books and papers on a wide range of topics in social theory and empirical sociology. He is co-editor of, and contributor to, *Ebbing of the Green Tide?: Environmentalism, Public Opinion and the Media in Australia* (1998). His interests currently centre on the links between scientific-technical change, cultural change and the ordering of social relations.

Astrid Gesche worked as a research fellow at the John Curtin School for Medical Research at the Australian National University before moving to Brisbane, where she completed a postgraduate fellowship in applied ethics at the Queensland University of Technology. She is now a member of the Centre for Applied Ethics at the School of Humanities and Social Science. Her main areas of research are the ethical, social, and legal implications of the Human Genome Project. In June 2000, she was invited to attend the prestigious ELSI Summer Faculty Program at Dartmouth College (USA). Astrid is a multi-disciplinarian and has published on topics related to medical history, eighteenth-century philosophy, applied ethics, and science. She is a member of the American Association for the Advancement of Science and other organisations.

Richard Hindmarsh lectures in nature, technology, and culture studies in the Contemporary Studies Program at the University of Queensland Ipswich. Richard is currently preparing a book about the social history of the recombinant-DNA controversy in Australia. He has three other research projects in train concerning genetic engineering: environmental ethics and regulation; regional perspectives; and holistic perceptions. Recent essays include *Consolidating Control: Plant Variety Rights, Genes and Seeds* (Journal of Australian Political Economy) and *The Problems of Genetic Engineering* (Peace Review).

Geoffrey Lawrence is Foundation Professor of Sociology and executive director of the Institute for Sustainable Regional Development at Central Queensland University. He has written widely about rural social issues in Australia and has conducted many national research projects, including an ARC-funded study of agro-biotechnologies. His books include *Capitalism and the Countryside* (1987), *Agriculture, Environment and Society* (1992), *The Environmental Imperative* (1995), *Environment, Society and Natural Resource Management* (2001), and *A Future for Regional Australia* (2001).

Rosaleen Love is a Melbourne writer on science and society. Her most recent book is *Reefscape: Reflections on the Great Barrier Reef*, on the cultural and scientific meanings of coral reefs (Allen and Unwin, Sydney, 2000 and Joseph Henry Press, USA, 2001). The US publication of *Reefscape* is simultaneously hardcover and electronic, at http://www.nap.edu. Rosaleen is senior research associate in the School of Literary, Visual and Performance Studies at Monash University, Clayton, Melbourne.

Janet Norton researched the social aspects of food biotechnologies in Australia as a doctoral student at Central Queensland University, receiving her doctorate in 2000. She was co-editor of the first edition of *Altered Genes* and has co-authored many articles on the sociology of food biotechnologies in Australia. Janet is currently working as a consultant social scientist, undertaking statistical research throughout Queensland.

Bob Phelps is director of the ACF GeneEthics Network, founded in 1988—a project of the Australian Conservation Foundation. Bob is an educator, environmental campaigner, policy analyst, and critic of new technologies, with 25 years' experience in the Australian and global environment movements. He promotes public understanding and debate on the environmental, social, and ethical impacts of gene technology. Bob participates in many Australian and overseas policy forums on

genetic engineering issues, including regulation, biosafety, food standards, sustainable systems, cloning, and the Human Genome. Bob is a
member of the Victoria University of Technology Institutional Biosafety
Committee and is registered with the federal department of industry and
science as a science and technology communicator.

Robyn Rowland, previous to her retirement, was head of the School of
Social Inquiry, director of the Australian Women's Studies Centre and
associate professor in women's studies at Deakin University, Victoria.
She researched the area of reproductive technology over eighteen years,
and has addressed meetings of MPs and State government committees
in Australia and overseas. She is the author of the book *Living
Laboratories: Women and Reproductive Technologies* (1992) and has
been the Australian and founding editor of *Issues in Reproductive and
Genetic Engineering: Journal of International Feminist Analysis,* and
Women's Studies International Forum. In 1996, Robyn was appointed as
an Officer in the Order of Australia for distinguished service in the areas
of women's health and higher education.

Anna Salleh is a science journalist with a special interest in science and
public policy. She has worked in print, radio, television, and on-line for
organisations such as the Australian Broadcasting Corporation and the
Australian Consumers' Association. Anna holds a BSc from Sydney
University, a Master of Arts (Journalism) from Sydney's University of
Technology, and is currently undertaking a PhD in science communication at the University of Wollongong.

David Suzuki is an award-winning scientist, environmentalist, and
broadcaster. Currently a professor at the University of British Columbia,
he is familiar to the public as the host of CBC's *Planet for the Taking*
and *The Nature of Things*. He is the author of 28 books, including
Introduction to Genetic Analysis. David is recognised as a world leader
in sustainable ecology, and is a recipient of Unesco's Kalinga Prize for
Science, the United Nations Environment Medal, and the Global 500.
He is a member of the Royal Society of Canada and the Order of
Canada.

Frank Vanclay is the Sub-Dean (Research) for the Faculty of Arts and
associate director of the Centre for Rural Social Research at Charles
Sturt University (Wagga Wagga, New South Wales). He has conducted

studies on social aspects of environmental management and social impact assessment. Books include: *The Environmental Imperative* (1995), *Environmental and Social Impact Assessment* (1995) and *Critical Landcare* (1997).

Tiffany White holds a masters degree in communication, and manages education and public affairs programs for the Sydney Catchment Authority. Her interest in biotechnology developed from studying—in a critical manner—the tone and frequency of science-based news stories, and the social implications of recombinant-DNA technology in the wider literature. She became concerned that both sides of the story about biotechnology were not being told, and concluded that the public was not developing a balanced understanding of biotechnology.

Peter R. Wills is a theoretical biologist and associate professor in the department of physics at the University of Auckland, New Zealand. His research interests include the molecular biological aetiology of spongiform encephalapothies, the thermodynamic basis of biochemical processes, and the physical origin of evolutionary self-organisation. He is known to the public as a critical analyst of the role of science and technology in society, especially military applications of nuclear technology and commercial applications of biotechnology. He has also engaged in academic work aimed at eliciting general acknowledgment of the legitimacy of non-scientific epistemologies, especially from indigenous Pacific cultures.

Acknowledgments

The first edition of *Altered Genes* sold out in 1999—only six months after its publication, and was nominated as one of the Australian *New Scientist*'s top-ten best-selling books on science for 1998-99. Our decision to produce this new edition arose from a genuine and continuing concern from our contributors that something was 'not right' with the way Australia was proceeding down the 'biotech path'. We thank our contributors for providing a sustained and coherent critique of biotechnology and genetic engineering. We nevertheless recognise that their views do not necessarily represent those of the organisations to which they are affiliated.

We also wish to express our appreciation to the original publishers of *Altered Genes*, Allen and Unwin, for agreeing to release the book to Scribe Publications. The proposal for a second edition was enthusiastically embraced by Henry Rosenbloom of Scribe, who believed that an updated version would provide new insights into the implications and impacts of genetic engineering in Australia.

We would like to express our appreciation to Janet Norton, co-editor of the first edition, for her careful editing and prudent judgement about content. Janét Moyle amd Emma Cotter provided valuable feedback on the original manuscript, while Sandra Henderson, Kees Hulsman, and Dimity Lawrence provided very useful comments on various chapters, as did Melissa Risely on the second edition chapter 'Constructing Bio-utopia'.

An Australian Research Council postdoctoral fellowship held by Richard Hindmarsh during the course of the preparation of the first edition—and support from the host facilities of the School of Australian Environmental Studies at Griffith University—was greatly appreciated. Similarly, research grants from the Australian Research Council and from Central Queensland University allowed Janet Norton to complete her doctoral studies and to be involved in the first edition of *Altered Genes*. The chapter by Robyn Rowland is an updated and amended version of Chapter 2 of her book *Living Laboratories: women and reproductive technologies*.

Richard Hindmarsh
Geoffrey Lawrence

Introduction

A geneticist's reflections on the new genetics

DAVID SUZUKI

I am delighted and honoured to have the opportunity to write an introduction for this book. Apart from having adopted Australia as my vicarious second home, I have found many colleagues there who share my interest in the social, economic, and political implications of the great human enterprise known as science. The kinds of ideas and considerations exposed in this book are at once focused on Australia, yet profoundly global in their significance. The issues raised here ring alarms just as relevant and important for Australia as for Canada and the rest of the world.

Discoveries and techniques for manipulating the molecules that specify the heredity and forms that all life takes represent truly revolutionary powers. The speed with which these capabilities have been acquired and the immense consequences of applying them demand widespread public examination and discussion. Having spent a quarter of a century as an active geneticist, and having been fortunate to have co-authored the most extensively used genetics textbook in the world.[1] I have nevertheless been concerned about the premature rush to exploit genetic technology—as well as the effusive announcements of its potential benefits. It may be instructive to indicate why I, a geneticist, have been led to raise questions and concerns about this exciting activity being carried out by my colleagues and students.

The human side of a geneticist

I was born on the west coast of Canada in Vancouver in 1936. I was the third generation of the Suzuki family in Canada and remember an early childhood full of happiness and promise. In 1942, my family was declared a danger to Canada, stripped of all rights of citizenship and incarcerated in a remote valley in the Rocky Mountains for three years. In 1945, we were presented with two choices: renounce our citizenship and leave the country or leave British Columbia for an exile east of the Rockies. The crime we, along with some 20 000 other Canadians, had committed was the possession of genes that had originated in Japan. The Japanese-Canadian experience during World War II is seared into my psyche, the source of immense insecurities and self-loathing. It instilled an excessive drive to gain approval from Canadians of my worth and it has made me hyper-sensitive to racism and injustice. Impoverished by the war and exiled in Ontario, my family seized on hard work and education as the means of improving our lot. As a recipient of a large scholarship, I was able to attend an outstanding liberal arts college in the US where, for an honours biology curriculum, I was required to take a course in genetics. Under the spell of my professor, I fell madly in love with this incredible subject as he posed fascinating questions and then revealed how scientists had uncovered the answers with precision and elegance. I was hooked and dreamed of going on to become a practising geneticist.

Sputnik & the explosion of science

The year before I graduated from college, the Soviet Union electrified the world with its dazzling achievement of sending Sputnik into space. In the agonising months that followed, American attempts to duplicate the feat failed spectacularly. In contrast, the Soviet juggernaut seemed invincible as they launched a series of firsts—dog, chimpanzee, man, team, woman—into orbit. It became clear that the Soviet Union was highly advanced in mathematics, engineering and science and, in response, the US poured billions into its space effort and into building up science. As a Canadian, I nonetheless became a beneficiary when I entered a US graduate school. It was a time of seemingly limitless opportunity for budding scientists. We were taught, and I believed, that science is the most powerful way of knowing, enabling us to examine and explain the deepest secrets of the cosmos. In our minds, nothing lay beyond the enquiring eye of scientists and with more research, life for all people would get better and better. I didn't question this article of faith and taught it to my students in my early years as a professor.

But two things happened after I graduated in 1961 as a fully licensed scientist that made me question my assumptions. When I began to teach, students would ask me questions that I couldn't answer; so I had to read more widely in order to find answers. In the process, I learned that there were parts of the history of genetics that none of my professors had taught me. And, as the years went by and the scientific community exploded in number and vigour, it became clear that despite the remarkable discoveries being made, life was not getting better for most people on earth while the planet itself was beginning to show unmistakable signs of distress. Let me discuss these points in more detail.

Genetic's dark history

As a teacher, I was enthralled by the rapid progress being made in genetics and loved to share the excitement by speculating on the benefits that could be gained from our insights. Students would often ask me questions which I found I couldn't always answer—like what about the moral and ethical implications of cloning, what is eugenics, who controls the application of genetic knowledge, what responsibilities does a practicing geneticist have, and so on. As I began to read, I discovered to my horror that, early in this century, geneticists, intoxicated with the powerful insights they were acquiring in this infant field, began to extrapolate from studies of physical traits in maize and fruitflies to behaviour and intelligence in people. In addition to promulgating the mistaken belief that such human attributes as drunkenness, sloth, criminality and deceit follow the same hereditary principles as feather pattern in chickens or seed colour in peas, geneticists began to write and speak of traits that were 'desirable' or 'undesirable', 'superior' or 'inferior'. But these are value judgements, not scientifically meaningful categories.

In their exuberance over the burgeoning field, eminent geneticists, not third-rate scientists, made bold claims that applications of principles from their discipline could improve the human condition. The consequences of their boasts were not trivial. Eugenics Acts were passed in many countries that resulted in sterilisation of people judged undesirable or inferior; American immigration restrictions were based on the notion that certain races have characteristics rendering them preferable, or less desirable; many states enacted anti-miscegenation laws to prohibit inter-racial marriage on the grounds that it leads to 'disharmonious combinations' in their children.

For me, the most shocking realisation was that the Holocaust was not the result of the ideas of a madman, but had been based on the grand claims of geneticists. After all, Germany had an outstanding community of geneticists who were at the cutting edge of the field, and many of them pressed for the improvement of Germany's gene pool by eliminating undesirables and encouraging the best. I learned to my horror that Josef Mengele, the infamous doctor at Auschwitz, was a human geneticist who held peer-reviewed grants for his studies of twins in the death camp. And in Canada, transcripts of parliamentary debates before and after Pearl Harbour show that politicians from British Columbia seized on the claims of geneticists as the rationale for eliminating part of the 'Yellow Peril'. The evacuation, incarceration and expulsion of Japanese-Canadians had been encouraged by the grandiose claims of geneticists! So two of the great passions in my life—the fight against bigotry and the study of genetics—intersected in a grotesque way.

What was most disturbing was that none of my professors had ever mentioned this part of the history of genetics. I had been taught science as a steady progression of insights by intellectual giants. It had never occurred to me that these heroic figures were fallible human beings, as competitive, jealous, ambitious and biased as any other group of experts. But equally disturbing was the fact that in many fields, but especially genetics, scientists in the 1960s and 1970s were making the same kinds of claims that history should have warned them against. Molecular biologists, understandably thrilled to have acquired incredible manipulative capabilities with DNA, were claiming to have their hands on the levers of life so that hunger, poverty, deleterious mutations, disease, and other afflictions were potentially controllable.

Sociology of genetics

The sociology of the scientific enterprise has changed radically since I graduated. In 1961, we were proud of doing basic research, seeking answers to questions that had no immediate application but added to our understanding of how life works. At conferences, scientists shared ideas openly and freely, co-operated on projects, respected areas already being investigated. There was excitement and a feeling of community. Today the incredible techniques for manipulating DNA have created a rush to patent DNA sequences and genetically altered life forms, and to realise massive fortunes. Where once scientists sought answers to fundamental questions, they now seek venture-capital investors for their biotech companies. Needing to produce ever more valuable output, researchers have become secretive, competitive and close-mouthed as the 'bottom line' has become driven by economics, not curiosity. In the 1920s and 1930s, scientists confused their personal values and beliefs about criminality, intelligence and abnormal behaviour with scientifically verified fact. Today, scientists fall into a similar trap when they confuse *correlation* with *causation*. Thus, as geneticists report on the correlation of a trait or characteristic with a specific locus or DNA sequence, they readily slip into an assumption of a causal relationship. This was striking when researchers reported an 80 per cent correlation of alcoholics and non-alcoholics with different alleles of the alcohol dehydrogenase gene. As soon as the report was made public, scientists and reporters alike discussed the allelic differences as the cause of the dependence or nondependence on alcohol. It's a serious mistake. It is like concluding that the high incidence of yellow-stained fingers and teeth among lung cancer patients proves that stained teeth and fingers cause the disease.

The human genome project & its real implications

The application of molecular techniques to the rapid sequencing of entire genomes is an astounding feat. I certainly never dreamed when I graduated that I would live to see the acquisition of such abilities in my lifetime. Biologists have designated the sequencing of the entire human genome with religious overtones. Nobel laureate Walter Gilbert, for example, declared that the project is the 'Holy Grail' of biology. The benefits of this project are widely claimed to be insights that will lead to understanding and cures for virtually every human condition for which there is a hereditary component.

The example of sickle-cell anemia should serve as a warning for those making such bold claims. The genetic and molecular determinants of sickle-cell anemia have been well known for over four decades but an effective treatment eludes us because of our vast ignorance about the human body, its components, and their functions. While the completion of the Human Genome Project will be a monumental technological achievement, it will merely provide a linear sequence of instructions that must be read and transformed into a three-dimensional entity that changes over time. We are a long way from knowing much about that.

As scientists exhaust the medical conditions for which sequence correlations can be screened, they will undoubtedly turn to other categories of human 'problems'—criminality, chronic poverty, alcoholism, homosexuality, intelligence levels, and so forth. And here, while the consequences of confusing correlation with causation are immense, the temptation to do so will also be strong. It seems far easier to blame the victim for his or her condition than to consider the impact of the social institutionalisation of racism, discrimination and inequities.

Technology—its benefits and costs

Biotechnology has vast social, economic and political ramifications. By focussing on this discipline, we gain insights into implications for technology as a whole. While scientists acquire knowledge about the world that is fragmented into bits and pieces, their discoveries can be applied with enormous impact. In chemistry, a host of new chemicals has been used for the production of synthetic materials, nerve gas and pesticides, for example. Discoveries in physics enabled the creation of X-ray machines, nuclear weapons and plants, radar, radio telescopes and compact discs. In medicine, researchers have provided antibiotics, nuclear imaging and oral contraception. The benefits have been immediate and obvious because we deliberately design technologies to perform specific functions. However, because our knowledge of the real world in which those applications will be used is so fragmented and incomplete, we have no capacity to predict beyond their immediate utility to their long-term consequences.

We have had enough experience with technology to know that, however beneficial, every technology, from bows and arrows to antibiotics, nevertheless has costs. But our ignorance ensures that there will always be deleterious effects that cannot be predicted beforehand. Take, for example, pesticides. When DDT was shown to be a powerful insecticide, the benefits were immediate and obvious—pests such as mosquitoes and weevils could be killed, and farmers and chemical companies would make money. At the time, geneticists already knew that resistant mutants would be rapidly selected so that more or different chemicals would soon have to be applied as the pests became resistant. And ecologists could have pointed out that insects are the most numerous group of animals on earth and are an essential part of all ecosystems. Only one or two of every thousand insect species are pests to humans and it is folly to spray broad-spectrum insecticides simply to eliminate the small number of pests. That would be like killing everyone in New York city to manage crime in the city! But no one could have predicted that when a molecule like DDT is sprayed at concentrations of parts per million, the molecules would be consumed and concentrated up the food chain by several orders of magnitude. That's because biomagnification was unknown to scientists as a phenomenon and was only discovered when bird populations were reduced significantly.

Over and over again, our intoxication with new technologies has been dampened by the discovery of far-reaching, unanticipated effects. Biotechnology will be no different but, because we are manipulating the very blueprint of living things, the consequences will be monumental. However, those who express concerns about the hazards of widespread use of new techniques or products are placed in the impossible position of having to define what the dangers might be. Because we know so little, it is not possible to know what deleterious effects there will be, even though we can say with certainty there will be hazards. Unfortunately, our technological activities are 'managed' on the assumption that unless a concrete hazard can be specified, one does not exist. It is that very way of thinking that has thrust us into a global ecological crisis.

Scientism

The remarkable success of science when applied by the military, industry and medicine explains the pre-eminence that the scientific community has acquired in this century. There is a widespread belief that science is the most profound way of knowing and allows us to know the 'true' state of the world. 'Scientism' is the belief in the authority of science and is widely held not only by scientists, politicians and businesspeople, but the general public. It is a dangerous myth, and scientists themselves should be the first to know and demystify it. Consider this.

I graduated as a fully licensed geneticist in 1961, full of excitement at the revolution in genetic thinking that was then going on. I was enthralled and arrogant. I felt scientists were pushing back the curtains of ignorance, and were understanding the deepest mysteries of biology. Today, when I tell students what we believed about the structure of chromosomes, how genes were organised and the mechanisms of development and differentiation, they laugh in disbelief. Considered today, in 2001, the leading ideas of 1961 seem naive and absurd. But students are then stunned when I remind them that twenty years from now, when *they* are professors, the hottest ideas of 2001 will seem as ridiculous to their students. Why then, I ask them, are you in such a hurry to apply the latest ideas? The nature of the scientific enterprise is that most of our current ideas will ultimately prove to be wrong, irrelevant or unimportant. This is not a denigration of scientists—science advances by having current ideas disproved. The very nature of the scientific enterprise undermines scientism as a belief. It is sobering to realise that as scientists rush to patent and capitalise on their incremental insights, the rationale for their applications will undoubtedly be wrong, wide of the mark, or trivial.

Science has occupied most of my adult life and has provided me with the happiest times and most exhilarating moments. I am proud to have been an active member of the scientific community and to have contributed in a small way to the body of scientific knowledge. It is because I love science and care passionately that I am compelled to share my concerns for the lack of understanding of the history of science, the nature of the scientific enterprise, how science differs from other ways of knowing, its power and its limits—and the responsibility of its practitioners. I am honoured to be part of this book which dares to reflect on the implications and the untrammelled benefits of the genetic revolution.

Part 1

Setting the agenda

1

Bio-utopia:
FutureNatural?

RICHARD HINDMARSH AND
GEOFFREY LAWRENCE

Genetic engineering confronts us with a new medium by which to imagine a future nature, one very different to the nature we have known for millennia. Some refer to genetic engineering as an 'inventionist ecology' for an invented landscape.[1] In a recent book called *Visions: How Science Will Revolutionize the 21ˢᵗ Century and Beyond*, technology guru Michio Kaku optimistically projects the futuristic vision of a biomolecular revolution with genetic engineering as its mainstay. Coupled to quantum theory and the computer revolution, genetic engineering will place us 'on the cusp of an epoch-making transition, from being *passive observers of Nature to being active choreographers of Nature*'. The Age of Discovery he declares is coming to a close, 'opening up an Age of Mastery',[2] where the biomolecular revolution 'will ultimately give us the nearly godlike ability to manipulate life almost at will'.[3]

Emerging in the early 1970s—after a gestation period of over forty years—genetic engineering is thus viewed by writers such as Kaku to be the most powerful tool yet developed for the manipulation of nature. Also known as recombinant-DNA (r-DNA) technology, genetic engineering blossoms from an age-old technology known as 'biotechnology'. 'Bio' is derived from the Greek term *bios*—meaning 'life'. People have for many thousands of years used varying technologies to modify, for human advantage, life forms and processes. Fermentation of beers and wines are but two obvious examples. Over time, new techniques for manipulating life have been added via industrial microbiology, biochemistry, chemical engineering and tissue culture.

Today, through genetic engineering, *modern* biotechnology (henceforth used interchangeably with biotechnology and gene technology) has become a massive corporate and state endeavour set potentially to make a major impact this century. The so-called 'bio-revolution' (after the industrial and informatics revolutions) is propelled by ambitious Genome Projects that entail the mapping of the genetic structures of humans, as well as those of commercially important plants, animals, and microscopic organisms. Once these worldwide projects are concluded, we may indeed enter the Age of Biology proper. The economic impact alone is deemed immense. According to bioindustry analysts, by the year 2025 some 70 per cent of the industrial economy and 40 per cent of the global economy will have, at its base, some form of biotechnology.[4]

The radical departure of r-DNA technology from previous biotechnical conquests of nature—such as artificial insemination, *in vitro* fertilisation, or plant breeding through traditional hybridisation—is illuminated by David Suzuki and Peter Knudston:

> In 1973, two decades after Watson and Crick published their revelations about the architecture of the DNA double helix, a historic event took place that marked the beginning of modern genetic engineering. Herbert Boyer, a researcher at the University of California, and Stanley Cohen, at Stanford University, succeeded in ferrying a recombinant DNA molecule containing DNA sequences from a toad and a bacterium into a living bacterial cell. There, to almost everyone's astonishment, the foreign toad DNA was copied and biologically expressed in protein. For the first time, scientists had choreographed genes from the cells of an evolutionarily advanced species to dance in the cells of a distantly related species.[5]

The techniques of genetic engineering thus lay open DNA (deoxyribonucleic acid—the carrier of genetic information) to human access and modification. Based on reweaving the threads of life through cellular mechanics, the so-called 'blueprint of life' can be redesigned at the submicroscopic 'inside'. Bioscientists can 'unzip' DNA and inactivate or delete it from cells, or isolate and recombine it in transgenic or cross-species transfers, producing novel or organisms hitherto unknown in nature. If the social power and influence of gene-altering proponents allows them to bioindustrialise the world, as they seem poised to do, dramatic change may follow:

> genetically modified crops, livestock and pastures will be leading contributors to the economy ... Gene and embryo technologies will tailor animals to their environment as well as to highly demanding markets ... deliver designer enzymes to create high-value rural products ... yield superior breeds of leisure animals, from aquarium fish to racehorses ... give rise to modified animals whose organs can be transplanted into humans ... Medicine will be revolutionised ... gene tests will determine the likelihood of an individual developing diseases and identify them, as well as behavioural and physical tendencies. Recreation will be enhanced by parks, golf courses and gardens containing wholly new organisms—plants with long life, unique colours and grasses that don't need watering or mowing ... Huge off-shore cages will be built for the rearing of fish ... that grow faster ....[6]

These are features of a bio-utopian future championed by scientists from Australia's Commonwealth Scientific and Industrial Research Organisation (CSIRO). 'In laboratories across the nation', they claim, 'the foundations of tomorrow are already being laid.'[7] With biological frontiers worldwide being recast by genetic engineering, these may be optimistic forecasts, but some are no longer situated in the realm of science fiction.

Mainstream media and biobiz websites depict heady, revolutionary images of the breakthroughs of the new biology—from scientific claims for 'big effect' genes that code for obesity, Alzheimer's disease, or for personality; to the engineering of plants to adapt to global warming and even for fanciful notions of converting noxious gases on Mars to create an atmosphere fit for human conquest and colonisation. Nothing seems impossible. The rate of genetics discoveries—which are fast outstripping other areas of scientific endeavour—parallels the zeal of biotechnology's proponents to promote their science as the key to human progress. Civilisation, finally, will be able to control its biological destiny. By escaping from their genetic straightjackets and the constraints of nature, people will find a new kind of freedom—a freedom to overcome disease and hunger, to have an improved standard of health, to ecologically restore the planet and, of course, to live longer.

Yet, sometimes developments from the laboratory make us sharply pause and deliberate upon the bio-utopian dream—and to ponder just what is going on. Is genetic engineering really rational science, or are the boundaries between reality and science fiction indeed becoming blurred and creating what critic Mae-Wan Ho—Professor of Biology at the UK's Open University—refers to as 'bad' science, or 'Frankenstein' science? Disinclined to share Michio Kaku's vision, Ho believes genetic engineering amounts to a worst case scenario of genetic determinism that 'offers a simplistic, reductionist description which is a travesty of the interdependence and complexity of organic reality', and which has the potential to destroy all life on earth. She argues that genetic engineering has no concept of the organism as a whole, nor of societies or ecosystems; that genetic engineering, at least in its current form, will not work; and that it is based on misconceptions that organisms are machines, and on a denial of the complexity and flexibility of the organic whole.[8]

Like Ho, an increasing number of people worldwide see Dr Frankenstein—the dominant image of the scientist in twentieth-century fiction and film—as what genetic engineering stands for. This is because of the enormity of the ethical dilemmas and inherent dangers of reconstructing nature through remoulding living substance. A central message of the author of Frankenstein, Romantic critic Mary Shelley, was that modern science—emerging from seventeenth-century Enlightenment Baconian, Cartesian, and Newtonian philosophy—was perceived to be separate from or 'external' to nature. It embraced the claim that science somehow possesses an 'objective', value-neutral quality that sets scientists apart from the rest of us by liberating them from moral deliberation or the consequences of their actions. Of course, the story of Frankenstein questioned this arguable quality by implying that scientists are not really *apart* from, but instead are *a part* of, nature and as such should be socially responsible for the consequences of their actions. This is vividly highlighted in the following passage:

> In scientific terms, the creation of the Monster is a brilliant achievement; yet Frankenstein's horror begins at the precise moment when the creature opens its eye, the moment when for the first time Frankenstein himself is no longer in control of his experiment. His creation is now autonomous and cannot be uncreated any more than the results of scientific research can be unlearned, or the contents of Pandora's box recaptured.[9]

A decade after Frankenstein was written, in 1828, the term *biotechnic* (meaning 'life' technology) emerged as symbolic confirmation of the western trend towards separateness from nature, whereby 'man [sic] has had to develop technology to make up for the loss of natural instincts'.[10] While both mechanistic and organicist versions of biotechnic visions emerged at that time, the mechanistic dominated in the new industrial age. A notable version acclaimed 'man' as *Homo faber*—'man the maker'.

For others, profound ethical dilemmas of genetic engineering are posed by H.G. Wells' novel *The Island of Dr Moreau*, with the extraordinary exploits of Moreau's vivisection of animals to remake them human-like. Yet, for others concerned about human experimentation, the new biology resembles that of Aldous Huxley's *Brave New World*, which describes a society in which humans are 'decanted' not born; where individuals are grown in bottles with the required characteristics to suit the life that they will be expected to lead 'as Alphas or Epsilons, as future sewage workers or [as] future ... Directors'.[11] And for those concerned about efforts to recapture the evolutionary past, *Jurassic Park* provokes images of unbounded nightmare. How far away is such a world? Huxley himself reflected: 'All things considered it looks as though Utopia were far closer to us than anyone ... could have imagined. Then, I projected it six hundred years into the future. Today it seems quite possible that the horror may be upon us within a single century'.[12]

His words seem prophetic. In 1997, a print-media article, breaking the story about the genetic production of a headless frog embryo from a UK laboratory, signalled just how close Bio-utopia may be:

> London: British scientists have created a frog embryo without a head, a technique that may lead to the production of headless human clones to grow organs and tissue for transplant ... such as hearts, kidneys and livers in an embryonic sack living in an artificial womb. Many scientists believe human cloning is inevitable following the birth of the sheep Dolly, the world's first cloned mammal ... [P]eople needing transplants could have organs "grown to order" from their own cloned cells.[13]

Yet, instead of addressing the issue of the morality of altering life that results in such bizarre and unnatural outcomes, the research was ethically justified as being in the services of humanity. After all, partial embryos—the result of the genetic manipulation of the egg to suppress development of the head—may not even technically qualify as embryos because they lack a brain or central nervous system. It was thus argued that the beauty of such developments was that because the donor is never sentient (or able to register pain) to begin with—'what could be the harm?'[14] If the morality question were indeed reduced to a mere technical consideration, the growing of partial embryos could potentially bypass legal restrictions and ethical concerns. Considering the moral outrage that the headless frog embryo instantly attracted—and before it the widespread concerns over 'Dolly' the cloned sheep and now over the increasing possibility of human clones, and their genetic enhancement as 'designer' babies—it is hard to imagine that only technical considerations will ever suffice. Indeed, such notions have attracted charges of being 'a violent assault on ethics'. As biologist Jacques Testart asks, 'Who can say in advance what is a good quality human species?' and 'Who should say in advance how the species should evolve?'[15] In a similar vein, Oxford University ethicist Professor Andrew Linzey argues that it is 'morally regressive to create a mutant form of life' such as the headless frog embryo. Indeed, to Linzey the work amounted to 'scientific fascism because we would be creating other beings whose very existence would be to serve the dominant group'.[16]

Such ethical concerns—as well as many others about manipulating DNA—are increasing both the intensity and scale of debate about the purpose and outcomes of genetic manipulation. Indeed, they highlight the dualism—the 'Janus-face' of biotechnology—which has emerged to focus vividly upon the perceived benefits and costs of genetic engineering:

> The genetic engineer, like a contemporary Daedalus, claims to be providing society with a vast range of innovations ... [Yet] as a result of the application of genetic engineering, the triggering of catastrophic ecological imbalances by the release of novel organisms into the environment, the creation of new agents of biological warfare and the increased power to manipulate and control people, may ... become realities in the near future.[17]

To its proponents, genetic engineering is a revolutionary scientific innovation that can provide new opportunities for improved health, protection from infection, control of diseases in crops, animals and humans, and for economic return. Some risk must be borne by society for technological progress. With a burgeoning world population, with increasing environmental pollution and social crisis, and with people's desire to live longer, the risk is worth taking. For antagonists, though, science is indeed proceeding down an unknown path into an anti-Utopia or dystopia where *algenists*[18]—latter-day alchemists—attempt to subordinate and reconstruct nature. The result could, perhaps, be ecological destruction on a scale never before seen. Given the visions and genetic recombinations now being concocted by genetic engineers, as well as our extremely limited knowledge about the workings of natural systems, there is little way of knowing what longer-term risks human beings and the wider ecosystem face.

Others are concerned about the social, political, and economic structures into which the new technology is embedded. As much of the new biology is driven by the motive of corporate profit, how much of the research is genuinely aimed at producing widespread and accessible community benefits? People fear that important questions of redesigning life and of playing God, of the appropriateness of the technology for an ecologically sustainable future, and of whose ultimate interests are being served will be swept aside in the race for commercial rewards. They argue—as was shown by the preceding Green Revolution of industrial agriculture[19]—that most of the benefits will go to those with power and wealth, not to those economically disadvantaged who are often touted as the eventual beneficiaries of the new Gene Revolution.

In response, proponents argue that a commercial edge is necessary for product development and for the widespread distribution of benefits from the new products: people risking money should rightly claim a return on their investments—something which, after all, will enhance their capacity to innovate for the future. The marketplace will determine what is developed, what produces profits, and what fails. Moreover the technology is viewed as being as safe as possible given available knowledge and the entrenched belief in science for progress. The counter arguments, however, include 'but, we have this area of great uncertainty'; 'that, given the dire environmental challenge of the twenty-first century, current models of science are not holistic enough to understand complex ecological systems'; and, 'but what about the broader values and needs of the wider community that differ to agree'. To sum up, while the proponents believe society must take a calculated risk with biotechnology, opponents believe that the risk is just too great.

Altered species being developed for the projected twenty-first century marketplace represent a cornucopia of novel medical, meat and dairy products, and fruit and vegetables in shapes, colours, flavours and textures never before seen. Already on the market, in the research and development (R&D) pipeline, or being talked about as future projects, are the oncomouse—a research animal with a cancer gene inserted into the animal's DNA; 'Dolly' the cloned sheep and 'Polly' the cloned lamb (with human-origin DNA inserted); edible plant vaccines; bioengineered plants that tolerate salinity, acidified soils, and resist disease; soybeans and many other crops tolerant to broad-spectrum herbicides; 'low mow' and novelty grasses (for example, luminescent and multicoloured grasses); non-germinating 'terminator' wheat and rice seeds; gene-altered animal organs (known as xenotransplants) designed for human transplantation; embryonic pig-human hybrids (possibly for therapeutic cloning); non-ripening 'anti-rot' tomatoes and pineapples; faster growing crops; biodegradable, plastic-producing, plants; violet carnations and blue roses; faster growing pigs and fish (some with human-origin DNA inserted); and cows encouraged to produce more milk via injectable r-DNA bovine growth hormone, also known as bovine somatotropin (bST).

New transgenic breeds of sheep, goats, and cows have been designed to secrete novel human proteins into their milk, later to be extracted and used in the pharmaceutical industry. The same animals can be programmed to produce nutraceuticals (higher-profit value-added food products) which—at least in theory—will result in healthier foods of greater nutrition. Rubber likewise is being developed to secrete pharmaceutical proteins. Other animals, also with human-origin DNA inserted, are redesigned to produce human blood plasma. These transgenic animals—designated mechanistically as 'bioreactors'—reside not on the farm but 'graze' at 'bio-pharming' companies such as Genzyme Transgenics Corporation. In December 2000, stocks rose 13 cents to US$2.50 for Viragen shares when the US company unveiled a genetically modified chicken designed to help develop drugs in their eggs, including monoclonal antibodies to fight cancer.[20]

The proponents' promise for these bio-products—feeding the world's hungry, improving the health of all, conquering disease, creating new 'clean and green' industries—is naturally alluring. Yet a darker scenario also exists. It is that of a bioindustrialised world dominated by a technical life-sciences elite (or bio-elite) whose collaborations or employment with corporations ensure profitmaking via patents on life and biodiversity—the new 'gene-gold', whose research concentrates on the symptoms of problems, not their causes. New food and health-care production systems based on a quick-fix approach will address the superficial 'needs' of a consumer society but fail generally to alleviate the growing social and ecological crisis which society currently faces, including the actual problem of consumerism now recognised as one of the world's worst causes of environmental degradation. Indeed, more directly, the ecological crisis may worsen through the widescale release of transgenic organisms into the environment, creating a new form of pollution—genetic pollution. This has already been signalled by occurrences of contamination of non-genetically engineered crops by gene-flow (or cross-pollination) from gene-altered crops planted in nearby or adjacent fields.

This Bio-utopia may also disadvantage society through increased social control, or through the genetic discrimination of neo-eugenics where children are genetically enhanced for a 'revamped' future free of pain, aggression, disease, or even sadness (as Rowland further comments upon in this volume). Will people be nominated as 'valids' or 'invalids' based on genetic status—as so enthrallingly depicted in the SciFi movie thriller *GATTACA*? We may come to live in a world where there is compulsory genetic testing and DNA fingerprinting, and where potential employers or insurance companies access human genetic information through DNA databanks. The latter was signalled during 2000 when Great Britain became the first country to allow insurers to use genetic tests to identify people with Huntington's disease. We may also see women's control over procreation further decreased by cloning and gene therapy, and by the promulgation of the myth of human perfectibility.

Bio-utopia may be a world where indigenous peoples' genes, and sacred knowledge and ownership of their flora's genetic information, are appropriated by Western firms for genetic databanks, and where potentially dangerous r-DNA vaccines are tested in weakly regulated countries. It may also be a world where horrifying new weapons of biological warfare emerge. As ex-bioweaponeer Ken Alibek revealed recently, scientists in the Soviet Union developed a genetically altered strain of *Bacillus anthracis* capable of resisting anthrax vaccines, and also a multi-drug resistant strain of glanders.[21] Closer to home, in early January 2001 *The Australian* reported that Australian scientists had accidentally created, through genetic engineering, a new virus that could destroy the immune system of mice. While the scientists said it was harmless to humans, they said it could nevertheless be used to create a genetically modified (GM) strain of smallpox which would render the smallpox vaccine much less effective.[22]

Farming and food manufacturing systems may also be altered in a manner that disadvantages—through price, accessibility, or lack of choice—those seeking choice, and the world's poor and hungry. Altered genes, in the form of novel foods, are already alienating the growing environmentally aware consumer movement. The movement is increasingly challenging the genetic manipulation of both nature and food, as seen in the recent collapse of markets for GM foods in the UK. As such, the agri-food industry—something in which people have tended, in past decades, to rely upon for the provision of healthy, nutritious foods—is now becoming a popular site of heightened struggle, not only over the 'meaning' of food, but also over its composition and delivery. This will continue to be the case where unlabelled GM foods continue to be sold to customers who believe consumer sovereignty has been violated, or who wish also to lodge their boycott vote against a *FutureNatural* of genetic engineering. Despite such unresolved questions about its promise, its ethical and scientific basis, its political alliances, and its social and environmental implications, the development of a global bioindustry continues unabated.

Biocolonising the global future

Global investment in biotechnology is now over US$20 billion each year, bankrolled predominantly by corporate businesses and governments in the industrialised world. While many companies, government agencies, and scientists aim to put molecular techniques to useful service, commercial imperatives shape and propel the technology. Handsome profits are envisaged. Global revenues of biotech products in 2000 were estimated to be US$15-25 billion, with human health care accounting for over 90 per cent of revenues. Top biotech drugs treat for red blood-cell enhancement, white blood-cell restoration, so-called 'growth failure in children', diabetes, prevention of hepatitis B, bone marrow transplantation, and multiple sclerosis.[23] Transgenic seeds have captured a major share of thr US acreage of cotton, soybeans, and corn, and plantings of GM varieties have grown rapidly in Argentina and Canada. Industry Canada estimates that the world market for biotechnology application will reach US$50 billion by 2005.[24]

Following the Boyer-Cohen recombinant-DNA breakthrough of 1973, hundreds of small bioengineering firms emerged in the US. Usually commercial 'spin-offs' from universities, they began to face corporate competition in the early 1980s as large companies such as Monsanto, Du Pont, Eli Lilly, Ciba-Geigy, and Merck began a catch-up operation to gain control of the emerging, lucrative market. Since then, takeovers, corporate mergers, and the vastly superior economic power of the corporate giants to invest in R&D and engage in expensive patent battles, government lobbying, and PR campaigns have acted to narrow the competitive field. With some 2800 biotechnology companies worldwide, transnational corporations now clearly dominate in controlling some 20–30 per cent of the fledgling bioindustry—a figure predicted to rise to 50 per cent in the near future.

In the top tier of life-sciences conglomerates are Aventis (formed in late 1998 through the merger of Germany's Hoechst and France's Rhône-Poulenc); Novartis (Switzerland) formed by the US$27 billion merger of Sandoz and Ciba-Geigy in 1996; and Syngenta—a recent merger of the agribusiness arms of AstraZeneca and Novartis, Monsanto—now a part of Pharmacia, Dupont, and Dow Chemical. Typically these conglomerates are a vertical and horizontal integration of many companies. For example, Novartis is the world's largest agrochemical company, the second-largest seed firm, the third-largest pharmaceutical firm and the fourth-largest veterinary medicine company. Recently, though, some setbacks to biocolonising the global future have emerged with the increasing global backlash against GM foods (as addressed below), with some of the large corporations divesting themselves of agbiotech divisions.

From the first field trials of a genetically engineered organism conducted in 1986, more than 4000 field tests of transgenic crops have occurred at more than 15 000 individual sites in 34 countries. Some 35 per cent (the majority) were for herbicide tolerance, including crops developed to tolerate or resist a particular company's own herbicide. Crops receiving the most attention include soybeans, canola, maize (corn), potatoes, melon/squash, and tomatoes, along with tobacco and cotton.[25] Australia's biodevelopment is intricately entwined with global biocolonisation.

Biocolonising Australia's future

Australia's place in global biocolonisation began at the 1975 Asilomar conference in California. The congress provided a platform for molecular scientists from all over the world, with Australian participation, to address the controversy that had emerged about the possible dangers of gene-splicing—especially those involving potential horizontal gene transfer of pathogenic material from the contained laboratory to the open ecosystem. The overall outcome was a defence of r-DNA experimentation and a rejection of any restrictive external interference in the development of bioscience (see chapter 2 for more detail). Asilomar, in short, provided a landmark opportunity to assert scientifically—and politically—the desirability of a global 'biofuture'.

Based on the not-inconsiderable intellectual resources within the CSIRO and the university sector, both bioscientific knowledge and the bioindustry have developed quickly in Australia. International connections have been enthusiastically sought as a means of both increasing funding for R&D and of gaining new and potentially lucrative world markets. By 2000, many of the 140 Australian firms involved in gene technology had formed strategic alliances with overseas firms, particularly from the US, Europe, and Japan. Another 30 firms were directly associated with large corporations, either as divisions or subsidiaries. Departing from this trend, in August 2000, Australia's 'flagship' biotech company CSL Ltd ventured overseas to clinch a A$1 billion deal for part of the Swiss Red Cross business in order to enter the lucrative US plasma products market.

The Australian Biotechnology Association estimates there are now many other Australian companies, some 1000 so-called 'minority or non-core users' that use some biotechnology techniques as part of their production and manufacture—in human therapeutics, food, veterinary medicine, agriculture, and other sectors. Other companies participate in the importation of GM foods from overseas where it is estimated that some 600 processed foods now contain GM additives and constituents, especially those derived from soya grown in the US.

Despite the commercial activity in Australia, gene technology has been used primarily within contained facilities, primarily laboratories of universities, hospitals, public research institutions such as the CSIRO, private or joint venture research facilities, and the commercial organisations referred to above.[26] At the time of writing, there have been eight applications for environmental or general (commercial) release of genetically modified organisms (GMOs). Three have been approved to date: Bt cotton; Melbourne-based company Florigene's violet carnation called 'Moondust'; and a carnation with improved vase life. Thirty to 35 per cent of the Australian cotton crop is now genetically engineered (see Salleh in this volume for a case study account of Bt cotton), and biotech giant Aventis is planning to have under cultivation a million hectares of its GM canola by 2005. This would constitute almost the entire Australian canola crop.[27]

Currently, the federal government invests some A$250 million a year into r-DNA technology, with the CSIRO capturing about one-third of those funds. New South Wales topped the list of state bio-industry competitive research funding with $139.8 million in 2000, Queensland attracted $101 million in Cooperative Research Centre funding, Victoria attracted $85 million in competitive research funding, and the other significant bio-research state, South Australia, attracted $47 million. Private investment exceeds $120 million. Australia's current market share, based largely on Bt cotton, amounts to about $200 million.[28]

In general, Australia's involvement in bioindustrialisation clearly indicates that the nation is increasingly dependent upon corporate decisions taken abroad, and based on the interests of distant corporate managers and shareholders. It is a position which has an element of neo-colonial dependency.[29] Australian involvement centres upon its roles as a specialist service-provider of scientific know-how to corporate interests; as a 'captive' technological client—both as adopter and developer—of the new biotechnological processes and products; as an active (if relatively minor) player and proponent in a global bio-policy network; and as a platform and route to the Asia–Pacific region for both local and global biodevelopment. Many top life-science firms have extended their operations into Australia and, like the situation overseas, a Byzantine web of formal contractual obligations and informal connections has emerged between public-sector research agencies—such as the CSIRO and universities, small biobusiness companies, and large global transnationals.

Public concern & social resistance

The failure to confront and resolve the broader environmental, ethical, and social implications of genetic engineering activities has engendered a global campaign of intense public criticism and protest. Worldwide, environmentalists, consumer groups, animal-rights activists, organic-agriculture advocates, food-trade organisations, and concerned scientists, feminists, farmers, religious groups, and indigenous peoples have all entered the debate. Almost without exception these groups have indicated how they (or their constituents) will be disadvantaged by application of the r-DNA technique.

Public concern has emerged in a diversity of colourful and sometimes provocative forms. Direct action has seen underground eco-activist groups like the 'Furious Viruses' and the 'Snarling Spuds' launch attacks during the late 1980s in the Netherlands and Germany on laboratories, hothouses, and test sites of r-DNA plants. More recently, so-called crop 'trashing' was a central tool used in the fight against GM foods in the UK. It has also spread to Ireland, to New Zealand, and recently to Australia where a group called 'Free Seed Liberation' raided a field of GM pineapples growing as part of a University of Queensland research project in Cleveland, a bayside suburb of Brisbane. Referred to as Australia's first known case of 'genetic terrorism', the anti-GM protestors uprooted a hundred plants reportedly valued at A$100 000.[30] Boosted by the news that taco shells across the US had been contaminated with GM Starlink corn not approved for human consumption, US activists have also begun to attack laboratories and research sites. The most serious was the torching of a research site belonging to Boise Cascade, the giant paper company which has been experimenting with GM trees.[31]

More peacefully, at the policy level, leading groups—such as the US Union of Concerned Scientists, the Third World Network, Green parties around the world, and the Australian GeneEthics Network—continue to lobby for strict regulation and attempt to raise public awareness and open debate. Internationally, many Green groups advocate a moratorium on environmental release of transgenic organisms, oppose herbicide-tolerant plants, and—together with a vast array of public interest groups worldwide—condemn the patenting of life-forms. A catalyst for the oppositional movement came in late 1996 when anti-genetic engineering group, the US Foundation of Economic Trends, joined with Greenpeace to initiate a global campaign to boycott Monsanto's 'Roundup Ready' soybeans. In very quick time the Pure Food campaign attracted support from more than 500 organisations in over 75 countries. A recent expression of this campaign was a 'People's Caravan 2000— Land and Food Without Poisons', consisting of thousands of peasant farmers, farm workers, and community groups marching across South-East Asia from India to the Philippines in protest against globalisation, pesticides, and genetic engineering.[32] This people's movement against genetic engineering confronts the rhetoric of proponents that biotechnology will feed the hungry of the world.

In response, a powerful proponent counter-campaign to align global society with a Bio-utopian future has been forged. Authoritative reports like *Agenda 21*—from the UN Conference on Environment and Development—support biotechnology as a core solution to the world's environmental and social problems. Paradoxically, it does so without addressing most issues raised by concerned third-world and environ-mental groups. Another oft-used tack is to depict critics as scare-mongers, extremists, zealots, neo-Luddites, as being unrepresen-tative of the wider public, or as working against economic progress. A 'softer' line is to suggest, in somewhat condescending fashion, that while criticism might be well-intentioned it is nevertheless ill-informed and wrong. In these ways, critics are construed as working against the interests of modern society. The public and critics are further told that they simply fail to understand the beneficial nature of genetic engi-neering, and that, if they could, they would see it as a proven, safe, and reliable technology. Consumers and critics are positioned, in short, to be scientifically naïve. The outcome sought is to marginalise their con-cerns outside the mainstream as well as to undermine public demands for participation in decision-making processes.

Most often, when criticism is levelled at new mega-technology, the result invariably has been a vigorous and uncompromising defence of the technology by its developers. The seemingly 'siege-mentality' of technical elites in the face of opposition is designed to place a fetter on discussion and debate. Rather than confronting public concerns about technology in a constructive manner, common approaches used to counter them are to marginalise them as we have discussed above, to outrightly dismiss them, or to *reassure* the policymakers and the public that nothing is untoward (as Crook elaborates upon in chapter 8). Such a response, though, is embedded in an even deeper malaise—one marked by the social insulation of scientific R&D and a technical reduc-tionism which drives much industrial innovation. Social and environmental problems then arise when socially insulated innovation is taken out of its vacuum and plunged into the bustle of the real world and confronted with all the diverse values of the populace.

Such problems are highly apparent with genetic engineering inno-vation. As a result, and in response, political uncertainty and public distrust has emerged in the form of conflict and social resistance. With many people bewildered by some of the emerging gene-altered life-forms, questions are being asked about just *what* is going on. Proponents are facing increasing difficulty in their attempts to dismiss or absorb mounting and accumulating public concerns. Enter the new weapon of the so-called 'public education' programmes. These have been developed to persuade a sceptical public of the virtues of genetic engi-neering (see also chapter 2 this volume). Recently, the US biotech industry united to launch a $50 million PR campaign to prevent a public backlash as has occurred elsewhere. A plethora of pro-biotechnology information packages is now visible on the Internet and in mailouts particularly targeted to younger, potentially more accepting, high-school audiences.

Public debate

In an open, democratic, society many people would challenge the limiting of public involvement in decision-making processes concerning the technological direction of society. Yet, as implied earlier, it is the view of many scientists and technical experts that the public is inexpert and its views sometimes alarmist. Many consider that public criticism and concern is thus unwarranted, that it restricts scientific freedom, threatens scientific 'neutrality', damages scientific and economic progress, and is a harbinger of the political control of science. What is denied by the experts is that science and technology in the modern world is increasingly regarded as a subset of industry R&D. In short, once considered separate entities, science and technology have now become predominantly 'technoscience'. Embedded in structures of subjectivity and social power, technoscience significantly shapes agendas for change, helps to direct society's preferences and determine economic and social outcomes, and has a major role in how we value and treat the environment. In this context, science is not neutral or apolitical, and scientists are not objective and 'independent' but are an inherent part of a system in which class, gender, racial, and other social relations affect— and are affected by—technoscientific change.

Some biotech proponents have accordingly responded to the debate over genetic engineering by proposing that matters scientific be removed from public debate. For example, genomic scientists Venter and Cohen proposed the establishment of a worldwide upper chamber of parliament to address the ethical issues of genetics research. The parliament would be 'a deliberative body of experienced scientists and philosophers ... to advise decision-makers in business and politics' and to 'inform the public of what is at stake in a given scientific advance and propose solutions'.[33] While this proposal might appear well-meaning it overlooks the point that scientists and philosophers bring with them particular views of the world—views which may not accord at all with those of the voting public. Where is the crucial role of the public's value systems in this top-down approach to directing technological change?

In democratic society an overwhelming case exists for public participation in matters impacting upon people's lives and on the biosphere. While pro-market technocratic advocates state this is the 'phoney democratic strand that suggests scientific progress should be determined by a community show of hands',[34] many good reasons exist for participatory technology. First, the public has a right to know exactly what is going on and to participate in major social and environmental developments—including the development and application of science and technology. Second, public participation is certainly called for when it is understood that, through its taxes, the public provides money for technoscientific experimentation and must also bear the consequences of its development. Third, effective representation of the public interest better assures the safety of commercial products and helps to anticipate—and therefore potentially to avoid—failures of science and technology. In other words, community input (both of knowledge and values) on technoscience and environmental issues can greatly enhance the quality of decision-making. The latter aspect is, of course, essential for a transition to ecologically sustainable development (ESD)—a national strategy that Australia has adopted to attain long-term social and economic development which sustains the natural environment and promotes social equity. Finally, public participation is warranted because society's traditionally perceived option of depending on the 'objective' analysis and expert opinion of scientists is compromised by their collaboration with the commercial and social power imperatives of (bio) technology. As is the case with most of the significant issues faced by society, public participation is essential if we are to avoid the potential abuses of power by those with vested interests.

These arguments require that other values—those of the public at large, those of local communities, those of indigenous peoples, those of groups and individuals affected by policy, and those of nonhuman species and of biodiversity—must be taken into account alongside considerations of economic benefit. It is no longer appropriate—if indeed it ever was—to leave the decisions to a technical elite, or to an amoral global marketplace. Participatory technology can help ensure that responsible scientific and technological decision-making occurs at all stages of technology choice and implementation.

At present, however, the public has a limited awareness of the broad social and environmental impacts and ethical implications of the so-called 'bio-revolution'. Yet this is a time in which Australians—as well as citizens overseas—are becoming increasingly concerned about genetic engineering and its potential effects. Worldwide opinion polls show that, whether people see benefits or costs associated with genetic engineering, most have moral reservations about tampering with life at the genetic level.

In response to the mounting public concerns about, and opposition to, genetic engineering—especially about GM foods—but also about genetic pollution, stronger sentiments have been expressed of late by some governments. During the year 2000, for example, the Tasmanian government imposed a one-year moratorium on field trials of GMOs while it assessed their safety in more detail; Western Australia also imposed a two-year freeze on commercial GMO releases; Australia introduced GM food-labelling laws, and in November brought into legislation the Gene Technology Bill 2000 which toughens up regulatory conditions; and the Queensland government introduced a draft ethics code of conduct for biotechnology R&D. Yet the latter two processes have been accused by critics of being biased to the voice of industry and as only endorsing token public involvement.[35] Voicing its displeasure with the Gene Technology Bill 2000, the Organic Federation of Australia said the Bill had done nothing to protect non-GMO and organic production systems from genetic contamination[36] (see chapters 2, 9, and 12 for further comment).

Overall, while the Tasmanian government and some governments in Europe especially are making serious attempts to grapple with the profound issues of gene technology, most—including Australia and the US—appear entrenched in a minimal effort to address the public's concerns or the need for thorough debate. They are ignoring calls by key communicators, public interest groups, educators, and consumers for informed, accessible, information. People are demanding comment on the broader issues—not only to enhance public awareness and participate in the debate, but *just to know what exactly is going on*. Currently, in Australia as well as elsewhere, little public information—except for the supportive accounts of proponents and government agencies—exists about the Janus-face of biotechnological change. In other words, at the very time that balanced knowledge needs to be promulgated, there exists a public information vacuum.

FutureNatural?: the purpose of *Altered Genes*

In the first edition of *Altered Genes* we presented a critical overview of the many important issues raised by modern biotechnology. In this book, released in 2001—the symbolic dawn of a new millennium—we attempt to go further. We believe that with the first wave of gene technologies and products now entering the marketplace, from the farm to the dinner table, there is urgent need to stimulate open and informed community participation in order to make appropriate decisions. If we are to realise an ecologically sustainable future, we must decide if and where biotechnology 'fits'. It is in everybody's interests to discuss and to decide upon the desirability of the genetic engineering enterprise and its Bio-utopian agenda. While the 'promise' of biotechnology has certainly been presented to the public, the broader issues and potential 'costs' remain under-presented. With social resistance to gene technology growing in contrast to state and industry PR campaigns promoting it, there is a need to strike a balance. The public needs to be drawn in from the 'outside' to become aware of the whole debate and be able to participate in an informed manner. As many of the contributors to this volume indicate, the public wants to increase its understanding and awareness of the issues that the new technology poses. To do so, the public must also recognise that gene-experimentation and commercialisation is embedded within the wider socio-economic, political, moral, cultural, and ecological conditions of contemporary society.

This book presents a range of issues that need to be brought to the public's attention. The viewpoints of its contributors, however, are by no means uniform in outlook. Some contributors see some promise—tempered by caution—in some of the applications of gene technology, but not in others. Other contributors find genetic engineering abhorrent, inherently flawed, or against nature, and think that the grand gene-altering experiment should cease immediately, or, that at the very least, a moratorium be imposed. Some believe it would be better to consider other, holistic approaches that work intrinsically *with* nature and community—natural therapies, deep ecology science, organic agriculture, and the like. Such approaches are rapidly gaining in popularity both overseas and here in Australia. Yet others believe the biotech baby should not necessarily be thrown out with the genetically engineered bathwater. Could there, for example, be something called 'sustainable biotechnology' which avoids transgenic options to concentrate upon genuinely advantageous bio-industrial processes? There are also contributors who display no strong position but raise important issues which society must face.

All contributors, though, would agree with two propositions: first, there needs to be thorough public debate; and, second, it is our right and duty as public academics and citizens to raise criticisms about—as well as to challenge the assumptions of—those preparing the world for a very different *FutureNatural*. The views of the contributors challenge the simplistic impressions that the lines of debate are clear and that the best way forward is to join one or other 'camp'—to adopt an 'us or them' approach. We recognise and applaud the diversity of opinion among the bio-critics, and we understand that there is—in parallel—(though with many covert) a diversity of views among the bioscientists and industry and government proponents. Through open community debate the critics, the proponents, and the public stand to gain a greater understanding, awareness, and appreciation of the issues and problems surrounding the move towards Bio-utopia.

The contributors' perspectives

The book has been divided into three sections, reflecting the major ideas and concerns of the contributors. *Part 1* provides a context for understanding and challenging the pervasiveness of biotechnology in public discourse and in practice, from the social to the ecological.

In chapter 2, **Richard Hindmarsh** traces the history and strategies of the proponent network for biotechnology which emerged in Australia from the early years of the 1970s. This network—with interlinkages of the CSIRO, universities, industry, government departments and various advisory bodies—continues today attempting to biocolonise the future. While DNA is consistently referred to as the building blocks of biotechnology, downplayed are the political manoeuvrings behind the scenes which are the other essential building blocks by which to construct Bio-utopia amidst dissent. Constructing the foundations for a *GMNatural* is followed at two 'fronts of action'—regulation control and information dissemination. Not seen in any simplistic conspiratorial manner, the campaign for biotechnology reflects an elite 'business-as-usual' and top-down technocratic approach to ensure unhampered progress for genetic engineering. Yet, despite the attempt to contain public debate and subdue knowledge about what exactly is going on, the oppositional movement continues to grow. In 'counter-attack', the bio-policy network has implemented a range of strategies that seek to 'enrol' public acceptance, and to block 'other' approaches to the future that may be more viable in a sustainable context. In this process, community concerns about genetic engineering have become submerged in an environment of weak regulations, disinformation, and complacent bio-elitism.

In chapter 3, **Peter Wills** questions the assumptions upon which the assessment, safety, and development of genetic engineering is based. He argues that modelling biological interactions as a linear process of genetic information—enshrined in the central dogma of molecular biology 'DNA makes RNA makes protein'—is a flawed representation of the relationship between genes and organisms. What is being missed is the fundamental circularity of genetic information processing, as well as the complexity of connections between biological entities and the environment. Overlooking these connections, molecular biologists and genetic engineers are showing themselves incapable of recognising the potentially dire impacts of the commercialisation, and rapid and large-scale dissemination, of genetically engineered products into the environment. If wide-scale release is allowed, then we can expect to per-turb the dynamics of genetic change to such an extent that the changing ecological patterns which have emerged, survived, declined, and re-emerged up until our current evolutionary epoch will be completely disrupted. Consequently, what is now urgently needed, Wills advocates, is an *indefinite* moratorium on the environmental release of transgenic organisms.

For **Tiffany White** the modern-day Lois Lane can never enjoy the journalistic licence of her fictitious counterpart from the Superman comics. The Lois of the real world would be hamstrung by deadlines, reliant upon the news and PR releases of companies, politicians, and sci-entists, and would not be encouraged to venture too far away from the newspaper's orthodox position on science and technology. The result, as White shows in chapter 4, is the propagation of pro-biotechnology information where the media act as legitimisers of current policy, failing to adequately incorporate alternative perspectives. The bioscientists and the biotechnology industry have a seemingly secure, authoritative, posi-tion in society. Their news releases are often designed so their content can be uncritically reproduced by journalists who are unwilling or unable to spend the time to 'test' their assertions or assumptions. In White's view, science journalists in the Australian commercial media need more support from their editors and news organisations if they are to succeed in the increasingly important task of engendering future public debate about important topics such as biotechnology.

Part 2 continues to explore issues of ethics of gene biology, and does so by focusing on important aspects relating to human gene map-ping, culture, and risk. In chapter 5, **Robyn Rowland** brings a feminist perspective to the topic of genetic engineering. Identifying the striving for 'perfection' as part of a narrow, masculine science, Rowland harks back to the eugenicists and their (misguided) attempts to create perfect human societies. Applied to human genome research, genetic engineering is viewed as a radical extension of male-controlled intervention in the lives of women. Women will be expected to conform to the notion of having 'perfect', 'unproblematic' children, and to carry genetically modified embryos through to term, to deliver the resulting children, and to rear them. Many interrelated issues are raised in this 'dream of quality-control' including the control of the technology in corporate and state hands, the patenting of DNA, genetic discrimination, and social control. To resist, challenge, or impede Bio-utopia from commodifying life and using women's bodies as 'living laboratories', the empowerment of women and the community in regulatory and decision-making processes is vital.

In chapter 6 the issue of genetic testing in the medical context is raised by **Astrid Gesche**. She summarises the problems associated with privacy and genetic information. Modern molecular genetics, she writes, allows us to test for both the presence of certain genetic aberrations and the likelihood of developing specific diseases in later life. But who, she asks, should have access to our personal genetic data, and what are the consequences of passing on this information? Should, for example, insurance companies or employers be afforded access to it? Gesche then proceeds to outline what privacy and confidentiality measures for molecular genetic information are currently offered by Australian institutions. She cautions that, at present, data and information derived from medical genetic testing cannot be safeguarded adequately. She concludes by suggesting that any genetic data transfer and exchange should be protected strongly by legislation. Without it, the potential for misuse is simply too great.

Rosaleen Love takes flight from the prevailing message that 'genes are destiny'. In chapter 7, she explores representations of the gene in the contemporary public culture of Australia, highlighting the so-called 'gene for death' and jokes about genes. By focusing upon the 'gene for death' she captures the paradox of the gene: that while life arises from genes, those genes also contain the potential for both deformity and death. Jokes about genes—a form of social resistance to genetic essentialism—provide a bridge between scientific and popular cultures, where non-scientists and scientists join in mocking current 'hyping' of the gene. Jokes point to the chinks in the 'world machine' view of life, the gap between the grandiose, hubristic claim that science has the answers and experience of the everyday world of increasing risk and hazard. A wider view of what it is to be human, perhaps a future view, is provided by the immunologist Miroslav Holub in his reflections on the human genome and the human spirit.

In chapter 8, **Stephen Crook** offers an analysis of biotechnology as 'cultural risk'. Crook argues that public anxieties about the risks of biotechnology are consistent with the pattern of an emergent 'risk society' as discussed by German sociologist Ulrich Beck. Biotechnology crosses fundamental boundaries between 'nature' and 'culture', threatening to pollute the sociocultural order just like witchcraft, or the eating of non-food animals. Risk-management is a major concern in risk societies, and Crook identifies 'regimes' through which biotechnology risks can be managed. Because biotechnology is culturally so risky, the most significant regime for its management is ritualistic and rhetorical. The proponents of biotechnology mobilise 'reassurance strategies' centred on rhetorical moves designed to show that biotechnology is 'natural', 'traditional', and safe rather than 'artificial', 'new', and polluting. Crook illustrates this argument with examples drawn primarily from debates about GM food. Although biotechnology's reassurance strategies are powerful, and resonate with dominant assessments of science, their victory however is not guaranteed. Anxious publics may distrust reassurances that they suspect are contrived solely to reassure.

Part 3 takes us to the farm, to the food table, and into the offices of those who are organising protests against a potential Australian GM *FutureNatural*.

In chapter 9, **Geoffrey Lawrence, Janet Norton, and Frank Vanclay** examine biotechnology's rise in the context of the public's desire to overcome environmental degradation, the farmer's need to become more efficient and productive, and the food industry's requirement to become more cost-effective and competitive. The 'biotechnological solution' is one which science has projected for all three. This chapter examines the claims of those supporting the current trajectory of agro-biotechnology research in Australia, and assesses the extent to which Australia's integration into the global food economy will be related to its utilisation of genetically engineered products. It is argued that while the state views biotechnology as a 'saviour' for struggling farmers—and more generally for an agri-food sector facing enormous economic problems—any promise is falling far short of the reality. There are concerns that farmers may not make predicted gains from their genetically manipulated inputs, that what constitutes a 'clean and green' agriculture may not equate with the novel products being developed by agribusiness, that release of genetically engineered organisms may lead to more environmental problems, and that a GM food industry may face consumer resistance—or outright rejection of many of the new products of biotechnology. There is simply no certainty that what bioscience is producing will be accepted by people—in Australia as well as abroad.

Anna Salleh provides a case study of one of the first genetically engineered inputs to agriculture, and the first GM crop in Australia— transgenic cotton. She explains, in chapter 10, how the soil microbe *Bacillus thuringiensis* (Bt)—used sparingly for many years by organic farmers as an environmentally responsible pesticide—has been appropriated by the biotech industry in an attempt to overcome pest destruction of cotton crops. The gene coding for the microbe's insecticidal toxin has been transferred into the cotton plant. In theory, this means the cotton plant will be able to protect itself from pests; if pests try to eat the plant, they die. Moreover, it is said, farmers are able to reduce the spraying of pesticides, making cotton growing a 'cleaner and greener' system of production. Salleh then poses some probing observations about the product. The more Bt cotton is produced the more likely it is that insects will develop resistance to it. What if the new biological insecticide is eventually rendered useless by the transgenic experiment? What will the cotton producers turn to then? What will happen to the organic farmers, no longer able to rely on Bt for safe emergency pest control? For Salleh, the potential over-exploitation of Bt threatens to see the gene 'worn out' in a very short time—rendering suspect the promise that biotechnology will bring 'cleaner and greener' agriculture.

In chapter 11, **Jean Christie** examines the geography and politics of biodiversity and biotechnology worldwide, and places Australia in this context. She points out that biodiversity, the raw resource for all biotechnologies, is mostly to be found in the gene-rich, but cash-poor countries of Africa, Asia, Latin America—and in Australia. This places Australia in a unique position: open for industrial exploitation like the developing countries of the 'South', yet politically in the camp of the industrialised 'North'. Christie contends that a desire by transnational industry to control genetic resources worldwide lies behind new industry trends, including bioprospecting for commercially important genes, and the escalating use of intellectual property rights to secure patent monopolies over genetic resources, and that two international agreements—the World Trade Organisation's intellectual property agreement, and the Biodiversity Convention—facilitate these trends. Using four case studies, she looks at the impact of these trends especially on indigenous peoples of Australia as well as those of the poor countries of the 'South'. Christie raises these concerns for heightened public awareness and debate, and for a better or more equitable outcome in the future for indigenous peoples and for the protection of biodiversity.

In the final chapter, **Bob Phelps** picks up the campaign trail in Australia. As director of the Australian GeneEthics Network—a key lobby group with the goal of attaining genuine public control over all policy aspects concerning genetic engineering—Phelps is well placed to relate the campaign struggle over genetic engineering. Topics he highlights include the lack of insurance confidence in the bioindustry, GM regulation, GM food standards and labelling, and the lack of adequate control over 'runaway' GM companies. Phelps provides a disturbing insight into the 'real' world of political manoeuvring being carried out to deploy gene-technologies without public knowledge or consent. He promotes the alternative of sustainable agricultural systems to that of genetic engineering. In highlighting citizens' campaigns against genetic engineering, globally and locally, he argues that, despite unbridled corporate power, people still have the opportunity to create a better legacy for future generations—a legacy that is free of a genetically engineered future.

We trust that these chapters provide readers with a variety of approaches to the topic of biotechnology and genetic engineering in society and environment. Our aim is to stimulate discussion about some of the most profound issues yet facing Australian and other societies and cultures around the globe.

2

Constructing Bio-utopia: laying foundations amidst dissent

RICHARD HINDMARSH

'Uproar over mutant meat' proclaimed a front-page article in Melbourne's *The Age* newspaper on 18 April 1990.[1] The genetics scandal exposed had been a cover-up for two years. The exposé not only flagged a possible and rather disturbing genetics future, but also challenged the long-standing claim that Australia's regulatory system for genetic engineering was a world-class model of scientific responsibility. 'Mutant meat' from some fifty transgenic pigs had been sold onto the Adelaide market by Metrotec—a joint venture between Metro Meats (Adsteam) and Bresatec (a University of Adelaide company)—without full authorisation from oversight committees and without public knowledge. A subsequent article, headlined 'The issue is the right to know'[2], captured a key and now-growing community and global issue concerning genetically modified (GM) foods.

'Metrotec failed in its duty to put the proposal to us', confirmed Professor Nancy Millis,[3] microbiologist and head of the federal government's then gene-technology supervisory body—the Genetic Manipulation Advisory Committee (GMAC). Despite this major breach of the regulatory guidelines, and ensuing calls by community organisations and Democrat senators for an inquiry to be held and for federal laws to be implemented for recombinant-DNA work, the government-funded research project continued without penalty. Funding could have been withdrawn, for instance, but was not. Critics charged this outcome as both irresponsible and inept, not only because regulators had consistently warned experimenters about breaches, and had gone to considerable lengths to set up an enabling regulatory system for optimal experimental conditions within the constraints of controversy about genetic engineering, but also because the University of Adelaide had been 'warned' over an earlier breach. Those responsible for that breach had also conducted a cover-up from 1984 to1986, and upon being detected had also suffered no penalty. Instead, a rather self-serving 'slap over the wrist' had been issued by regulatory head Millis for the prior misdemeanour. Millis' letter of censure to the University's Vice-Chancellor read in part:

> It is most regrettable that eminent researchers have violated the voluntary monitoring system ... There is genuine concern among the public and within the general scientific community over some aspects of recombinant-DNA research. This monitoring system was originally set up by researchers *to regulate themselves* in a manner which is responsible, generally effective and *minimises interference in research* [my emphasis]. However, any violation is likely to cause considerable disquiet in many quarters ... If incidents like this one were to become common then government may have to revise its attitude towards a voluntary system of monitoring.[4]

The latter was the last thing genetics experimenters would want to hear; a situation the in-house regulators knew and had actually campaigned against for many years. The threat of external monitoring thus posed a rather heavy deterrent but apparently not one heavy enough for the recalcitrant researchers at the University of Adelaide. This saga, of course, underlines the important problem of in-house or peer-review scientific regulatory control. Public trust and participation, scope of adequate assessment and risk, safety, sustainable futures, and honest and open accountability, are all at issue here.

In 1994, with those regulatory breaches and their associated issues seemingly long forgotten—at least by biotech proponents—readers of the *Canberra Times* were confronted with yet another rather odd headline 'Yes, we'll eat those tomatoes'.[5] A reported pilot survey on public attitudes to genetic engineering had informed the readers that, 'Most Canberrans would like to try eating genetically engineered tomatoes, and many would willingly grow them in their home gardens'; or so claimed the survey's sponsor, the Department of Industry, Science and Technology (DIST). Such boosterism was a misrepresentation of the information. It was one of the first in a questionable history of 'information' dissemination exercises deployed to win over the public to a Bio-utopian future. Despite the survey's fanciful claims, DIST declined to release for scrutiny its full findings to the *Canberra Times*,[6] or to bio-critical campaign group the Australian GeneEthics Network (whose campaign is outlined in chapter 12). Later, the main survey would show up highly questionable methodology and interpretation that would strongly contradict the department's claims.[7] Misinformation or disinformation, especially disseminated by public servants, is largely a hidden but another key issue in the debate over genetic engineering.

The disturbing events and issues raised so far however signal only the tip of an intense—though little known—debate occurring in Australia for nearly thirty years. Indeed, within Australian scientific circles it seems to have begun in about 1968. With molecular biologists 'cracking the DNA code', Macfarlane Burnet—Australia's greatest virus researcher—'warned that tinkering with the genes of bacteria and viruses was a dangerous pastime ...'[8]

Since the late 1980s, with many field trials of GM organisms conducted, and with GM soya and cotton grown in the US and the latter in Australia (see chapter 10), this debate has intensified quite dramatically with many environmental, consumer, and public interest organisations challenging the effort to biocolonise the future. The many issues raised though have been consistently marginalised. So how are stricter control of recombinant-DNA regulation and public knowledge and debate being stymied? Why are government agencies conducting politically sanctioned campaigns to educate the public to the 'benefits' of this potentially dangerous technology without adequate consideration or open community debate about its many profound issues? In short, why are the proponents getting away with it?

These questions are addressed here by considering how the social agenda behind the development and regulation of genetic engineering has been constructed or shaped to directions largely exclusive of public knowledge, debate, and participation. The analysis makes it quite clear that science and technology does not develop in a political and economic vacuum as a value-free, objective undertaking as 'science' would have us believe. Instead, it is embedded in existing economic and political relations—or, as some would say, in *social power* relations. Here, power is exerted by actors (or agents) through the exercise of influence and strategy (tactics, campaigns, forays, and countermeasures, for example) to secure favourable outcomes sought by those actors.

As the famous French analyst on *science-in-action* Bruno Latour tells us, 'Technoscience is war conducted by much the same means. Its object is domination, and its methods involve the mobilization of allies, their multiplication and their drilling, their strategic and forceful juxta-position to the enemy.'[9] In this apparent 'war of conquest', where science and technology has become the industrial medium for gaining social power and shaping society, 'actors work out their impulses to grow, to transform themselves from "micro-actors" to "macro-actors" by subduing others ...' In other words, scientific knowledge not only involves scientific inquiry through well-defined methods, but is signif-icantly influenced by the social construction of that knowledge through value definition, through negotiation, through enrolment of allies to par-ticular views, and through the blocking of other views.

In relation to genetics, two key 'fronts' in the seemingly 'techno-scientific war' to construct a Bio-utopia are those of *regulation control* and *information dissemination*. Through in-house regulatory control, proponents actively 'organise off' the regulatory policy agenda many ethical and ecological issues associated with gene-technology proposals, as well as social ones such as the consequences of the technology's application on people's living and working conditions, or the implica-tions for indigenous peoples of the acquisition of nature through 'bioprospecting' and patents (see chapter 11). While gene-technology proponents consider agenda-fixing tactics as essential for biocolonising the future, others view them as suppressing equally important issues and other modes of production they consider more viable for a sustain-able future, such as organic agriculture. Shaping the policy agenda in this way is known as the 'mobilisation of bias'.[10] Mobilising bias by fixing regulatory policy to predominantly a genetics 'technical' basis makes it much easier for genetic engineering to proceed, amidst dissent.

Facilitating and legitimating in-house regulatory control is the strategy of disseminating information or images that project sanitised and favourable aspects of the r-DNA technique and that downplay, ignore, or trivialise its unpleasant aspects. Herman and Chomsky would refer to this process as 'the manufacturing of consent' or as 'the creation of necessary illusions'.[11] In the classic study of Australian propaganda, Alex Carey would define it as 'setting the terms of debate' or 'managing public opinion' and, within the industrial context, as 'protecting corporate power against democracy'.[12]

The capacity of bioproponents to undertake such manoeuvres, and which aids them to secure enormous R&D funding, is visibly strengthened by their location in existing dominant structures of influence in the policy terrain of Australia—including the scientific establishment—represented here, for example, by the Australian Academy of Science and the Commonwealth Scientific and Industrial Research Organisation (CSIRO); by industry bodies like the corporate-dominated Australian Food and Grocery Counci,l and the Australian Biotechnology Association; and by government agencies such as Biotechnology Australia.

The following account of *bioscience-in-action* begins at the regulation 'front'.

Setting the regulatory agenda

Following the remarkable recombination of toad and bacterial DNA in 1973 by US scientists Boyer and Cohen, some molecular biologists and US environmental groups became increasingly agitated about possible biohazards arising from shifting genes across species barriers. Noted US biochemist Paul Berg, prompted by concerned reaction to his own research of inserting the tumour-producing virus SV40 into *Escherichia coli* (*E.coli*)—a bacterium commonly found in the human gut—began also to have reservations. Moreover he was receiving disturbing daily phone calls from fellow scientists: 'They'd ask "Send us pSCIOI [a variety of DNA]". We'd say "What do you want to do?" And we'd get a description of some kind horror experiment and you'd ask the person whether in fact he'd thought about it and you would find that he hadn't really thought about it at all.'[13]

The likelihood of public controversy led, in February 1975, to the US 'Who's Who' of molecular biology issuing a 'call to arms' to colleagues to air a proposed safeguard of significant implication to the emerging r-DNA research community—a voluntary moratorium on hazardous r-DNA experimentation. The 'Asilomar' conference (in California) brought together 140 molecular geneticists, microbiologists, and biochemists, particularly 'veterans' from the influential national academies of science of industrialised countries, including two sent by the Australian Academy of Science.

The outcome sought was to reassure the scientific community and the general public that gene-splicing could be done safely (the strategy of 'reassurance' is elaborated upon by Crook in this volume).[14] What amounted to a two-fold 'pincer-movement' strategy was quickly deployed by the bio-elite. The first movement quelled internal scientific disagreement arising at Asilomar about the narrow scope of biohazards considered. Scientists were encouraged to compromise and to ensure a consensus of scientific opinion about a projected low risk of bio-experimentation. The second movement interpreted this 'scientific review' to the wider community as social responsibility, and thereby projected 'authoritatively' that the scientists could voluntarily self-regulate. The message relayed was: 'The technology is safe, we are responsible and wise, and we need open flexibility to experiment and develop'. Compulsory regulation of r-DNA work was thus presented as an unnecessary waste of time and public resources, and any ideas of a moratorium were swept aside.

On the basis of the Asilomar findings, the US National Institutes of Health (NIH) developed what became international guidelines for contained r-DNA experimentation. In-house scientific oversight committees to administer these guidelines were subsequently instituted internationally by the emerging global alliance of bioscientists. The enabling regulatory foundations for colonising a bio-future were being installed.

In accordance with this strategy, the Australian Academy of Science—the elite representative of Australian Science—formed its Committee on Recombinant-DNA Molecules (ASCORD). Facilitating the proponent's plans was Australia's science policy, embedded in a non-interventionist market approach. This afforded ASCORD the unique opportunity to shape r-DNA monitoring processes from the ground-up with minimal interference from any government wishes to do otherwise. An immediate manoeuvre was to restrict membership of ASCORD to a scientific alliance of Academy fellows, CSIRO scientists, and university colleagues. This would not only serve to shape Australian regulation to that of the international bioscientific community, but also to inhibit internal dissent. With top players in the field controlling the regulatory committee for government funded proposals, who in the bioscientific community applying for funding would want it widely known that they might morally disapprove of any particular biotech R&D directions? ASCORD thus gained ascendancy to allow r-DNA experimentation to occur along the politicised and limited safety lines established at Asilomar.

Outflanking community concern

Despite such moves to contain controversy, public concerns soon arose. An unconvinced Sir Mark Oliphant—a noted physicist and Governor of South Australia—warned that genetic scientists were waiting to start research that 'could produce uncontrollable epidemics', and that there was an increasing call worldwide to ban such experiments.[15] In 1977, an ABC *Four Corners* programme and the print media probed further into the international controversy about safety as well as into Australia's measures to ensure safety.

In response—and briefed earlier by ASCORD personnel—science minister Webster issued forth in the Senate that there was 'no risk to public health in those experiments presently carried out in relation to recombinant DNA research in Australia'.[16] Yet Department of Health (DoH) bureaucrats thought otherwise. Fearful of possible disaster, DoH officials thought that, should it occur, government could better handle the repercussions than could ASCORD. Through internal negotiation, the two parties soon reached an agreed position for proposed cabinet submissions: r-DNA supervision should shift to a government-convened committee, and reliance should be retained on ASCORD personnel and procedures.

Dissent, though, still existed within the bureaucracy. Sir Hugh Ennor—the-then science department secretary—queried the proposed ASCORD submission over its failure to canvass the desirability of stimulating public debate. His counterpart in the health department replied:

> It was my impression that your Minister concurred with the view that a formal public inquiry was not warranted … he took the initiative in arranging for an informal presentation of the issues by Professor Ada to Parliamentarians as a substitute … I understand the reactions of the Parliamentarians were quite moderate … there are dangers in fostering additional public debate … better decisions will be reached if the matter is not carried to the extremes reached in the USA.[17]

Professor Ada, the ASCORD chairman, though, had presented genetic engineering as only a 'little different' from the norm. Soon after this exchange, Ennor's views were moderated to an incoming science department secretary by bureaucrats supportive of biotech. They highlighted a general easing of the overseas controversy about genetic engineering, but without offering an explanation that this had occurred after an intense mobilisation of bio-elites internationally to assist the US bio-elite, who had been unable to consolidate the US self-regulatory front amidst enduring public dissent over safety considerations.

At a series of unannounced and private meetings in the US and Europe in the mid-1970s that were restricted to a select 'phalanx' of bioscientists, a familial pattern emerged. With leading bio-elites enveloping the meetings in a 'state of siege' atmosphere—us against them, science against 'antiscience', proponents against critics, biodevelopment against research annihilation, and so forth,[18] internal dissent and contrary views were organised off the agenda. As had occurred at Asilomar, the agenda became further politicised by introducing restrictive assumptions concerning *degrees* of safety about r-DNA experimentation rather than a commitment to safety *per se*. Analysis was limited to a subset of biohazards. A central assumption made was that *all* r-DNA research was conducted with *E.coli* strain K12—even though NIH guidelines catered for other host organisms. Important issues about *E.coli* K12—such as low-level seepage of novel gene-combinations from research sites into the environment—were submerged, and the hypothesis that the scientists constructed was 'that whatever else might be done to it, it was impossible to convert *E.coli* K12 into an epidemic pathogen which could escape the laboratory and run rampant through a population'. Known as the 'epidemic-pathogen hypothesis', it was afforded celebrity status in the early campaign for legitimising biotechnology.[19]

In Australia, r-DNA operatives had also become increasingly complacent in their bias. For example, at Australia's first general scientific meeting considering r-DNA technology in 1977, dissent had surfaced when Professor John Walsh—a University of NSW geneticist and dean of the Faculty of Medicine—suggested that precautionary action to control any hazards of r-DNA experimentation should be considered in the form of external legislation. His suggestion was met by a 'deafening silence', and further organised off the conference agenda by ASCORD members in attendance who dismissed public concerns about r-DNA experimentation.[20]

The first major skirmish in Australia between community and r-DNA interests then erupted. When the Walter and Eliza Hall Institute of Medical Research began constructing a C3—containment of high risk—laboratory at nearby Parkville for r-DNA experiments, the University of Melbourne Assembly launched a high-profile public inquiry in 1977. *The Age* immediately issued an article called 'Lab horrors feared'. In voicing its protest, the Assembly exposed the logical paradox within the scientists' standard defence of their research: If experimentation was so safe, then why were such elaborate containment procedures required? After a two-year long inquiry the Assembly's report was published in 1979—which, incidentally, was the same year that the CSIRO declared biotechnology as a priority research area. The report declared little

Yet, though Australia had raised its banner in the early 1980s to capture emerging biobiz (as chapter 1 details), powerful reservations were held in the bureaucracy to loosen Treasury and Finance purse-strings and change the non-interventionist science policy to actively support bioindustrial development. Subsequently, the Labor Opposition raised high-technology development as a key issue in the 1983 federal election. Since that election, which Labor won, Australian federal and most State governments have enthusiastically supported biotechnology.

Regulatory skirmishes renewed

Commercialisation, especially for agriculture, involved the release into the open environment of genetically engineered organisms. Proposals for field release first emerged in the USA in the early-to-mid 1980s. A significant controversy paralleled this about the desirability of reconstructing nature genetically and about the limitations of existing regulation to assess the risks of field release. Ecological disaster was feared. Campaigns against r-DNA technology were renewed.

Despite the international controversy, Australian scientists developed a proposal for environmental release in Australia. It was to field-test a live bacterial vaccine in animals. This was the catalyst for the RDMC scientists to tell Barry Jones—the then minister for science and technology—that they lacked expertise to adequately assess the risks of release of transgenic organisms into the environment. Yet, while they reasserted their role as the 'rightful' guardians of genetics regulation, they offered no suggestions as to how to address the broader *ecological* aspects. At the same time, Sydney University biologist Ditta Bartels raised the social tenure of their work, which also challenged the wisdom of genetics-focused scientists (with interests in the technology) dominating the regulatory committee:

> the recent Australian proposal ... should have provided a unique opportunity to engage the community in a dialogue concerning the new directions in which recombinant DNA work is proceeding. But instead ... there has been no disclosure ... *Commercial-in-Confidence* has been stamped on both the proposal and the documents relating to its assessment ... all details ... are kept securely behind the closed doors of the RDMC.[23]

To make matters worse for the bio-alliance, the environment department then began to launch 'forays' to 'capture' the RDMC on conflict of interest grounds. In questioning the Department of Science and Technology's role in regulation, it exposed the fact that it was not normal practice within the government bureaucracy for the monitoring or regulation of potentially hazardous activities to lie with agencies responsible for their development. The environment department's position was boosted by support from environmental heavyweights, the Australian Environment Council and the Australian Conservation Foundation (ACF). All environment bodies wanted more ecological assessment and representation on the RDMC.

To counter this new opposition, the bio-policy network set in motion a series of counter-manoeuvres to demonstrate that the bioscientists were 'thoroughly in control' of the technology, and that they could be *seen* to be exercising that control responsibly through being situated in a *neutral* bureaucratic setting (yet well out of range of the 'greenies'). A two-fold adjustment to regulatory policy was effected following the defeat of the environment department in an intense bureaucratic battle. First, the RDMC was transferred to the Department of Administrative Services and renamed the Genetic Manipulation Advisory Committee (GMAC). Second, its scope of experts was expanded slightly to include (limited) ecological input. These moves 'neutralised' the environment department and temporarily absorbed public pressure, which was then further contained by a consensus view reached within the bureaucracy that it was *not* now necessary to initiate public debate about environmental release.

Undoubtedly, the federal government's prioritising of biotechnology that year (1987) had had a significant influence on events. Corporate biobusiness had emerged solidly overseas. This posed new opportunities to check the decline of market capitalisation for biodevelopment under the impact of the stock-market crash. Commercial incentive was then furthered through more relaxation of the GM regulatory guidelines. Interest in Australian biodevelopment was subsequently renewed. Transnational companies such as Groupe Limagrain and ICI began to invest heavily, and some fifty Australian biotechnology companies and research institutes renewed collaborative R&D agreements with foreign companies.

Public protest, though, did not decline. More community groups entered the fray, and the ACF raised the stakes by demanding a five-year moratorium be placed on all environmental releases of genetically engineered organisms. At about this time the 'Uproar over mutant meat' exposé surfaced in Adelaide.[24] Environmental groups bayed: 'An inquiry is needed into the secret releases and the attempted coverup'.[25]

In June 1990, industry and technology minister Senator Button sent the terms of reference for an inquiry on genetic engineering to the House of Representatives Standing Committee for Industry, Science and Technology. Some 35 per cent of the 167 submissions—those of the public—demanded an immediate halt to genetic engineering work in Australia.[26] Public dissent, however, went unmentioned in the report's tabling to parliament. This reinforced the inquiry's bias in supporting genetic engineering: 'The Committee believes that the possible economic, environmental and health benefits derived from genetic manipulation techniques are worth pursuing, even if not all of the claimed benefits materialise'.[27] For the internationally competitive environment sought, any role for public participation in regulation could only be a token one.[28] Yet, to enforce a regulatory position that would both facilitate biodevelopment throughout Australia and contain any 'rogue' r-DNA experimenters, a shift to a mandatory and harmonised regulatory system was also canvassed. Protest was thus 'absorbed', and oppositional groups retreated. Like the Australian Academy of Science report of 1980, the 1992 inquiry report was then used as an authoritative and 'persuasive' vehicle to convince decision-makers and a sceptical public of the benefits and 'low risk' of genetic engineering, and to legitimise these views as central to r-DNA policy development.

Manoeuvres at the dawn of the twenty-first century

Eight months after the inquiry report was tabled in 1992, the federal government formed a consultative group (generally supportive of biotech) to negotiate the proposed shift to a mandatory harmonised State–Commonwealth regulatory framework. Yet, it took another five years just for the federal government to endorse the inquiry's proposed statutory legislation. During that time there were many internal disagreements in Commonwealth-State negotiations. But, with industry becoming more ready to commercialise biotech products on a large scale, regulatory negotiations were revived in earnest in 1997. Finally, the Gene Technology Bill 2000 was tabled in the House of Representatives for debate.

From 1992 to 2000, the bioscientist-dominated GMAC continued to administer the regulatory regime. Field trials increased, especially for herbicide-tolerant crops. Approval was forthcoming for the importation into Australia of Monsanto's 'Roundup Ready' soy beans despite well-informed public dissent. Commercial planting of GM cotton was also approved by the National Registration Authority despite many criticisms (see Salleh's account in chapter 10), and another four apparently deliberate breaches of the guidelines occurred—including the dumping of GM canola plants at an open commercial tip at Mount Gambier in South Australia in March 2000, and the inadvertent mixing of 69 tonnes of GM cotton seed with conventional or non-GM seed in Queensland. Despite adverse publicity about these latter events, the year 2000 saw another favourable outcome for the bioindustry with the abandonment by State and Territory health ministers of an agreement for a tough labelling scheme for all GM food contents, due to intense pressure from the federal government and industry (see also chapters 9 and 12).

Information dissemination

Parallel to, and supportive of, self-regulatory manoeuvres, information dissemination exercises designed to manufacture public consent for genetic engineering began to emerge more visibly. Following the 'misinformation' lead of the Australian Academy of Science report of 1980, the Australian Biotechnology Association—with a membership of business interests, scientists, and government officials— began in 1990 to disseminate to secondary schools a free information-pamphlet series presenting an informative though overly rosy account of biotechnology.

This effort was followed-up by the CSIRO. In 1990, its top officials believed that the organisation's future depended upon the widescale release of genetically engineered organisms. Yet they had also recognised this as a highly 'sensitive' community issue. A major strategy was hatched to calm the waters for the proposed GM future.[29] The CSIRO timed its entry into the pulp bio-information market during the latter stages of the parliamentary inquiry. It launched a $250 000 'Genetic Engineering: Will Pigs Fly?' travelling exhibition across Australia which clearly sought to enrol public acquiescence to genetic engineering.[30]

In 1995 another two organisations shored up the information front. The first—the Public Liaison Committee of the regulator GMAC—began to disseminate 'educative' pro-biotech brochures to secondary schools. The second was an elite PR unit called the 'Gene Technology Information Unit' (GTIU) which, holding a two-year $500 000 contract from the Department of Industry, Science and Technology, emerged more professionally to generate a favourable image of genetic engineering to science teachers and their students—the most favoured target of an international effort to secure public acceptance.[31] The GTIU's flagship—a glossy brochure called 'Gene Technology at Work'—painted a winning picture of the views of biotechnology proponents while trivialising those of critics (see also Crook's account in this volume of the use of rhetoric by bioproponents).

More recently, as the battle over GM foods intensified, the federal government implemented a multimillion-dollar National Biotechnology Strategy. This began in May 1999 with the creation of a new agency, Biotechnology Australia, in the Department of Industry, Science and Resources. Its basic mission is to support biotechnology development as a key source of future economic growth. Initially resourced with $17.5 million it received another $31 million in the May 2000 federal budget for the following four years.[32] An important part of its mission is to provide 'balanced' information to create public 'awareness' of the benefits and issues of biotechnology. To achieve its goal it runs rural forums, conducts a phone-in gene technology information service, and disseminates a biotech information booklet for secondary school students, a glossy leaflet on GM foods for Australian supermarkets, and so-called fact sheets. Overall, the images conveyed are those of soft sell propaganda to support pro-biotech public relations.

Dissension

Despite such manoeuvres by the powerfully resourced pro-biotech coalition, the oppositional movement has maintained its critical edge against genetic engineering as part of an escalating global movement which has enjoyed some success. Recent and intense opposition in the UK over environmental release and unlabelled GM foods has seen markets suffer badly for GM foods, as they have across Europe. This has prompted the European Union—or at least some member States—to adopt a *de facto* moratorium on the approval of new genetically modified organisms (GMOs), pending a review of current legislation. Proposed changes include increased public involvement in the regulatory process, and improved monitoring of GMOs for release. Brazil—Latin America's biggest agriculture producer—adopted a more oppositional stance and banned GM foods in July 2000, and Ireland has launched an organic development program.

In Australia, Tasmania became one of the first states in the world to declare GM crops a forbidden pest, and in July 2000 imposed a one-year moratorium against their trial.[33] The State's Primary Industries Minister, Mr Llewellyn declared a 'safety-first policy' saying, 'We need to be certain GMOs won't pose a risk to our health, or our environment, or our agriculture'. This announcement followed in the wake of genetic contamination that had occurred through cross-pollination from GM crops to non-GM crops overseas, and the recent Australian breaches described above. New Zealand also succumbed to the pressure of public outcry over genetic engineering. In April 2000 the New Zealand government announced a twelve-month Royal Commission on genetic modification. Pending its outcome, New Zealand instituted a voluntary moratorium with regard to field tests and open release of GMOs.

The Gene Technology Bill 2000

In the meantime, following Australia's Commonwealth-State regulatory negotiation process where consultations with major stakeholders included industry but excluded community groups, and where it was agreed that the regulatory 'burden' be kept to a minimum, the long awaited Gene Technology Bill 2000—introduced into parliament on 22 June—was enacted in early December with all manner of clauses and provisions. Critics promptly accused the federal government and opposition of making a 'backroom deal' that ignored most of the 'healthy' recommendations from the lead-in Senate report. According to the Organic Federation of Australia (OFA), the amendments did nothing to protect non-GMO and organic production systems from contamination through cross-pollination and horizontal gene transfer. Bob Phelps, the Australian Conservation Foundation GeneEthics convenor, stated that the 'gene safety laws ... are as useless as a flyscreen in a cyclone' (see also Phelps in this volume). The Senate report—*A Cautionary Tale: Fish Don't Lay Tomatoes*—had proposed several key recommendations including that the new Office of the Gene Technology Regulator be a statutory authority rather than a single individual; that communities be told the location of released GMOs; and that those licensed to grow GM crops not contaminate non-GM produce or land.

Yet in the amended Bill communities may be denied knowledge of the location of GM crops or field trials; those who damage or hinder authorised GM activities can suffer two-year jail terms and hefty fines; and the regulator will not be a statutory authority and will need only to receive one stream of expert advice on each and every GM application. This advice would come from the scientific committee (in all probability, another rehash of GMAC and its predecessors the RDMC and, before it, ASCORD). The recommendation that the Gene Technology Community Consultative Group (with community interests represented) would also have this function was not followed through in the amendments. In other words, the community committee—rather than being potentially a transformational mechanism to protect community interests—can be seen as a legitimising device of co-option, enabling proponents to claim that the regulatory system has been overhauled when, in fact, central power remains with the scientists and, by default, with bioindustry. The government and opposition jointly voted down amendments to ban members of the regulator from having links with commercial activities relating directly to gene technologies. Significantly, there is no right of appeal for third parties to the Administrative Appeals Tribunal against any decision to issue a GMO licensed by the regulator, which would have facilitated accountability. The only recourse for a third party will be via the federal courts—a very expensive and time-consuming process, and therefore largely inaccessible to public and community interests. 'This Bill is fundamentally going to affect every farmer and every consumer, and people should be very upset by this deal', commented the OFA's Scott Kinnear.[34]

Conclusion

Our trip to the regulatory-control and information-dissemination 'fronts' in the 'technoscientific war' to construct the foundations for Bio-utopia has revealed that an environment of weak regulations and lack of information, indeed disinformation, has been crafted in Australia—despite a massive and enduring escalation worldwide of important community concerns about the implications and impacts of genetic engineering, and increasing calls for public participation, open knowledge, and debate. Clearly, in-house regulatory control, biased public education or awareness programmes, and contained public debate are central avenues for manufacturing consent and subduing dissent for the proposed Bio-utopia. To achieve this, the bio-elites reside strategically as an influential network of allies—recombining behind the closed doors of increasingly complex and consistently stacked regulatory systems, of myopic and self-interested industry associations, and of government bureaucracies unrepresentative of the general public. As a result, the spectre of nature and society being biotechnically reconstructed to an invented commercial DNA landscape is now looming on the horizon.

In 1979, before genetic engineering became entrenched, the University of Melbourne Assembly asked, why, if recombinant-DNA experimentation was so safe, were such elaborate containment procedures required? Now, at the dawn of the twenty-first century, after nearly three decades of experimentation, research, and development, and with widescale commercial releases of GM organisms imminent, the same question should be posed in slightly different form. If the technology is so safe and predictable, and offers environmental sustainability as its proponents are so keen to recount, why are the elaborate and complex regulatory structures dictated by the Gene Technology Act 2000 needed? Surely, if biotechnology were working *with* nature in an interconnected way these measures would not be necessary? Moreover the exclusionist manoeuvres of the bio-elite do not invite public trust and participation, nor honest and open accountability. One must ask: is this really the way to an ecologically sustainable future? Is this how we want the future to be—*FutureNatural* becoming *GMNatural?* Or, in dissent, would we rather prefer *EcoNatural*—that is, Nature as is?

3

Disrupting evolution: biotechnology's real result

PETER R. WILLS

ny form of engineering is an attempt to manipulate the world for some desired end. Genetic engineering is no exception. It manipulates DNA molecules which carry hereditary information needed to determine the biology of individual organisms. The concept of genetic information as the real cause of biological phenomena—ranging from enzyme structures, through disease processes to the character of ecosystems—has grown up as molecular biology has advanced during the last 60 years or so. This belief in genetic determinism is now fostered quite uncritically in the public perception especially by proponents of biotechnology.

In undertaking a critique of genetic engineering and its likely consequences, we must therefore look into the scientific assumptions, conceptions, and interpretations of molecular biology—the underpinning science of genetic engineering. This chapter begins with an overview of some of the basic ideas of molecular biology and then proceeds with a discussion of the relationship between genes and organisms. With this background, we attain a position to make a sensible assessment of the long-term consequences of the genetic engineering currently being undertaken.

All scientific enterprise is based on implicit assumptions which in effect define the 'paradigm', model or framework of thinking that the practitioners of particular disciplines of science use. The paradigm of molecular biology and genetic engineering is termed 'reductionism'— a belief that all phenomena find their ultimate explanation in terms of elementary physico-chemical processes and events occurring at the level of atoms and molecules. Starting around the late 1930s, advancing techniques in physics and chemistry started to be applied to the detailed microscopic analysis of intracellular structures and processes.[1] Molecular biologists now wish us to believe that our understanding of the molecular processes that occur within cells, and what this knowledge tells us about organisms, has been determined without any recourse to ideas beyond those which have been proven to be 'true' in physics and chemistry.

Although the reduction of biology to physics and chemistry is something which can in no sense be proven, it is so widely assumed in scientific circles to be an obvious fact that it even forms the foundation of the public regulatory procedures used to evaluate genetic engineering projects (see chapter 2 for a sociopolitical analysis of this aspect). The dominant reductionist paradigm of molecular biology has been challenged by the work of theoretical biologists investigating the relationship between microscopic physical processes and the phenomena we observe, especially biological phenomena, in our everyday 'macroscopic' (visible) experience.[2] Even when microscopic entities follow very simple rules, interactions among populations of them can give rise to enormously complex so-called 'emergent' macroscopic phenomena.[3]

Emergent properties are so characteristic of biology and so pervasive in the living world that the task of actually finding explanations based only on explicit molecular descriptions looks impossible—in principle as well as in practice. Seemingly external environmental factors can serve as constraints that have enormous influence on the overall mode of behaviour of complex biological systems, to the extent that internal molecular workings often do not appear particularly relevant to what is observed. For example, the extinction of the large blue butterfly—*Maculina arion*—has been attributed to the introduction of myxoma virus into rabbits in Britain, not because of any direct molecular effect of the virus on butterflies, but because the reduction in rabbit numbers changed the ecology of the entire system on which the butterflies were dependent for survival.

The reductionist viewpoint then becomes more like an article of faith: the laws of nature and order in the universe are those of physics and chemistry alone; the ultimate reality is comprised of the symmetries and properties of the unified field describing the Big Bang and the subsequent quantum mechanical unfolding of the material universe, including all of biological evolution.[4] By looking at the character of this reductionist assumption we can unravel the issue of how genetic engineering is likely to affect the ecological and evolutionary processes that define life in the global biosphere.

Molecular biology & reductionism

Knowledge about the molecular mechanisms operating inside cells has increased rapidly since the explanation of the basic modes whereby genetic information is transferred from one material form to another. The simplest, most evident path of genetic information transfer is summed up in what Francis Crick dubbed 'The Central Dogma' of molecular biology: 'DNA makes RNA makes protein'.[5] DNA (deoxyribonucleic acid) is identified as the genetic material which carries the blueprint of the cell. RNA (ribonucleic acid) constitutes the copies which are made from the cell's DNA library and transported from the nucleus into the cytoplasm and elsewhere for use in maintaining cellular functions. The proteins inside cells are catalysts—which carry out the vital metabolic functions that keep a cell going—constructed from the information in the RNA. The Central Dogma contains another 'truism' or obvious truth for the molecular biologist: 'Once information has got into protein it cannot get out again'.[6] That is, information cannot be fed *back* along the pathway—protein to RNA to DNA. Interpreting biology in these terms is an easy proposition if one adopts a reductionist attitude. That attitude can be characterised as reductionist on two counts. First, there is the assumption of *material* reductionism: 'Now that we know the physical structure of the genetic material, we may state the aim of molecular biology as defining the complexity of living organisms in terms of the properties of their constituent molecules.'[7]

It is assumed that what we observe comprises a set of material constituents—atoms, molecules, and so on—and that change occurs via interactions among these elementary constituents. Although many individual biologists hold personal beliefs in something 'beyond' what can be discovered by using the methods of physics and chemistry, material reductionism is carefully maintained as the framework for intellectual discourse in molecular biology.

The second assumption is that of *genetic* reductionism—that all of the significant physico-chemical features that distinguish organisms are explainable in terms of their genetic constitution. The strongest proponent of this idea is Richard Dawkins, for whom the whole of biology is a mathematical 'space' of organisms represented as nothing but genetic sequences:

> Sitting somewhere in this huge mathematical space are humans and
> hyenas, amoebas and aardvarks, flatworms and squids, dodos and
> dinosaurs. In theory, if we were skilled enough at genetic
> engineering, we could move from any point in animal space to any
> other point. From any starting point we could move through the maze
> in such a way as to recreate the dodo, the tyrannosaur and tribolites.[8]

All biological phenomena—ranging from the short-timescale internal
workings of individual cells all the way up to the broad sweep of evolution over the billions of years since the appearance of the genetic
code—are thus nothing but a complex game among replicating DNA
sequences that compete for control of the resources needed to reproduce. Material and genetic reductionism are assumed to be 'proven'
because they seemingly explain why Darwin's principle of natural selection is correct and why its competing theory—the Lamarckian
conception of evolution through the inheritance of acquired characteristics—is incorrect. Whereas Lamarck argued that evolutionary progress
occurred because organisms somehow passed onto their progeny biological improvements they experienced in their lifetime, Darwin
proposed that evolution was simply a matter of the 'survival of the
fittest'—the characteristics of organisms are determined by the genetic
material they inherit, and changes occur only as a matter of the random
mutation and recombination of parental genes.[9]

Yet molecular biology differs from physics and chemistry, upon
which it is supposedly based. Whereas much of physics is concerned
with abstract theoretical problems and the interpretation of basic concepts—such as questions of the type: 'Is quantum mechanical
indeterminacy a true feature of the world or just a reflection of our
incomplete knowledge?'—molecular biologists act as if crystal-clear
answers have already been given to questions concerning the kind of
reality they are dealing with. All biological explanations, however, are
couched in language which refers—either explicitly or implicitly—to
functions which can only be defined contextually, usually in relation to
whole organisms which cannot be specified completely in terms of their
molecular parts.

For example, the protein which functions as the oestrogen receptor
on the surface of a cell is only called by that name because other cells
produce oestrogen molecules which migrate and bind to those particular
proteins on the cell surface. There is nothing intrinsic to any protein
molecules which define them as 'oestrogen receptors'. It may be

assumed that descriptions of biological phenomena are reducible to descriptions in terms of the most basic concepts of physics and chemistry; but this cannot be proven, and the topic is seldom discussed by biologists.[10] Yet, while molecular biologists have been developing techniques of gene splicing, *theoretical* biologists have been busy re-analysing the concepts on which biology is based, and have—by-and-large—conceptualised biology as something more complicated than simple reductionism would indicate. Most of this theoretical discussion is lost on those who practice molecular biology or genetic engineering because they are seemingly not interested in the interpretation of what they do; they are more concerned with the manipulation of nature than with its understanding. It is our task to make the relevance of these discussions clear within the context of evaluating genetic engineering, so we'll begin with the concept of a 'gene'.

The character of genes

If we start with the molecular biological description of a gene—that is, a DNA sequence which codes for the amino-acid sequence of a protein—we discover very quickly that a 'gene' is a much more complex entity than is often apparent. A description of a particular human gene is given in Figure 3.1, but this coding sequence is only part of the DNA which gets transcribed into the messenger RNA to be translated by the ribosomes in making the protein product. [Ribosomes are the microscopic particles that synthesise proteins inside cells by reading the genetic messages that have been transcribed from DNA genes into messenger RNA. Ribosomes themselves contain some fifty different proteins.] The whole gene actually consists of three parts found at different locations in the cells' DNA, and the RNA copies of these three parts have to be spliced together to give the 'mature message' which the ribosomes use. The ribosomes need more than just the coding sequence. At a bare minimum, the mature RNA must have an extra sequence before the coding region that tells the ribosome where to start translating the message into a sequence of amino acids to make the protein. There are other parts of the RNA which are left over from the splicing process. Other non-coding parts of the message probably function as signals recognised by other proteins which exert control over ribosomal function. Thus, even as a DNA sequence, a full gene is much more than the coded specification of a protein sequence.

We are used to thinking of genes in relation to inheritance, and that

Figure 3.1 **Sequences of the human PRNP gene and prion protein**

The upper line of text (lower case letters) specifies the normal human PRNP gene (Kretzschmar *et al.* 1986) in terms of the four nucleotide bases adenine (a), cytosine (c), guanine (g) and thymine (t) that are joined to form the DNA chain. The lower line of text (capital letters) specifies the translation of the gene into a sequence of 240 amino acids, abbreviated to the standard one letter code (Lewin 1994, p. 11), that are joined in sequence to form molecules of the prion protein (PrP). Positions of two variations in the DNA sequence are indicated. In about one-third of the human population a 'gtg' codon specifying 'V' instead of an 'atg' codon specifying 'M' occurs at position 129 in the protein sequence. A mutation from 'gac' and 'D' to 'aac' and 'N' at position 178 is associated with inherited Creutzfeldt-Jakob Disease (in the case of 'V' at position 129) or otherwise Fatal Familial Insomnia (in the case of 'M' at position 129).

```
atggcgaaccttggctgctggatgctggttctctttgtggccacatggagtgacctgggc
M   A   N   L   G   C   W   M   L   V   L   F   V   A   T   W   S   D   L   G

ctctgcaagaagcgcccgaagcctggaggatggaacactgggggcagccgatacccgggg
L   C   K   K   R   P   K   P   G   G   W   N   T   G   G   S   R   Y   P   G

cagggcagccctggaggcaaccgctacccacctcagggcggtggtggctgggggcagcct
Q   G   S   P   G   G   N   R   Y   P   P   Q   G   G   G   G   W   G   Q   P

catggtggtggctgggggcagcctcatggtggtggctggggggcagccccatggtggtggc
H   G   G   G   W   G   Q   P   H   G   G   G   W   G   Q   P   H   G   G   G

tggggacagcctcatggtggtggctggggtcaaggaggtggcacccacagtcagtggaac
W   G   Q   P   H   G   G   G   W   G   Q   G   G   G   T   H   S   Q   W   N

aagccgagtaagccaaaaaccaacatgaagcacatggctggtgctgcagcagctggggca
K   P   S   K   P   K   T   N   M   K   H   M   A   G   A   A   A   A   G   A

                                 g
gtggtgggggggccttggcggctacatgctgggaagtgccatgagcaggcccatcatacat
V   V   G   G   L   G   G   Y   M   L   G   S   A   M   S   R   P   I   I   H

                                 V
ttcggcagtgactatgaggaccgttactatcgtgaaaacatgcaccgttaccccaaccaa
F   G   S   D   Y   E   D   R   Y   Y   R   E   N   M   H   R   Y   P   N   Q

                                 a
gtgtactacaggcccatggatgagtacagcaaccagaacaactttgtgcacgactgcgtc
V   Y   Y   R   P   M   D   E   Y   S   N   Q   N   N   F   V   H   D   C   V

                                              N
aatatcacaatcaagcagcacacggtcaccacaaccaccaaggggggagaacttcaccgag
N   I   T   I   K   Q   H   T   V   T   T   T   T   K   G   E   N   F   T   E

accgacgttaagatgatggagcgcgtggttgagcagatgtgtatcacccagtacgagagg
T   D   V   K   M   M   E   R   V   V   E   Q   M   C   I   T   Q   Y   E   R

gaatctcaggcctattaccagagaggatcgagcatggtcctcttctcctctccacctgtg
E   S   Q   A   Y   Y   Q   R   G   S   S   M   V   L   F   S   S   P   P   V
```

a gene in some way 'stands for' some feature of an organism so that genetic variation is related to variation in the characteristics of organisms within a population. For example, within the human population, there is a reasonably homogeneous distribution of individuals who carry a PRNP gene which encodes the amino acid valine (about one-third of the population) instead of methionine (about two-thirds of the population) at position 129 (see Figure 3.1). Based on studies of mice which lack the prion protein and which seem to have disturbed diurnal rhythms, we might speculate that the particular version of the PRNP gene that we carry (or combination of versions—we each have two copies of this gene, one inherited from each parent) exerts some minor influence on our sleep patterns, but nothing obvious has yet been found.

Other variations (mutations) in the PRNP gene are associated with inherited neurological disease, and it has been found that the detailed pathology of the disease, its timecourse, and the types of spongiform lesions seen in the brain on autopsy are characteristic of each mutation. So far, the idea of genes being carriers of individual biological traits seems to hold up. However, there is one example of a mutation in PRNP which demonstrates that the situation is not quite so straightforward. An aspartic acid to asparagine mutation (at position 178) gives rise to Fatal Familial Insomnia in carriers of the amino acid methionine (at position 129) and familial Creutzfeldt-Jakob disease in carriers of the amino acid valine. Thus, the type of disease caused by a mutation is modulated or altered by what would otherwise seem to be an entirely neutral feature of another part of the gene, demonstrating that the genetic encoding of any individual characteristic is inseparably entwined with the coding of others, even within a single gene.

We have now discovered what is perhaps the most important feature of genes and is most overlooked by molecular biologists: the biological meaning of genetic information is context-dependent—each part depends on all the others, just as the words in a dictionary, looked at in purely linguistic terms, are all defined in terms of one another. No gene in and of itself is the carrier of a single isolated trait. Any trait, like the occurrences of any word in a dictionary, is in fact a set of relationships between a large number of other traits, all of which coexist in a relationship of *interdependence*. This does not mean to say that there is no pattern to the effect of similar genetic changes in different organisms. In fact, many genetic diseases in humans—including many cancers— can be mimicked in mice by finding mutations linked to the human disease and introducing virtually identical mutations into the genes of mice. It is actually this cross-species regularity in the way genetic changes are manifest as traits in organisms—from mice to humans—and our partial knowledge of the general mechanisms of genetic expression, which make the whole enterprise of genetic engineering possible. But we would be foolish to think we could understand all the complexities of genetics on this basis.

The processes that occur inside cells are linked together in complicated hierarchies, making the effects of changes in genes non-additive, whether the change is made to a single base, a whole gene, or perhaps some associated DNA control element. In other words, because genes act at many levels, the effects of simultaneous changes in two or more genes are not usually the same as if the changes were independent. And, if a genetic engineer takes a piece of DNA from one organism and puts it into another, the overall effect of the transformation depends on the location of introduction of the new DNA relative to genes already there. The products of many genes control the expression of others, often in complex cascades of 'signalling' processes involving the production of proteins that bind to specific DNA sequences and regulate their expression, like the *tat* protein that promotes the transcription of HIV genes in infected cells. The effect of these regulatory proteins is often modulated by their interactions with other molecules present in the cell. In spite of this complexity, the view persists that the effects of altering genes can be understood as if they were somehow additive, as if the effects of simultaneous changes in different genes could simply be added together, irrespective of any complicated interactions.

There is a deeper reason why the linear model of genetic information processing enshrined in the Central Dogma ('DNA makes RNA makes protein'—assumed adequate for the purposes of genetic engineering) provides only an incomplete representation of the relationship between genes and organisms. A sort of chicken-and-egg dilemma exists with respect to the genetic code. Genetic information can only be decoded (or translated into protein) in a cell which contains ribosomes able to translate genetic information, and ribosomes are themselves largely products of the translation process. And where do cells obtain their initial sets of ribosomes? They inherit them from parent cells. A cell which inherited no ribosomes could not decode any genetic information and would die. The decoding of genetic information is therefore not a linear process—it is circular—a closed loop which has been sustained unbroken in every lineage of biology, stretching back, in some cases, many trillions of cellular generations. DNA can only be thought of as a blueprint for an organism because the organism's vital processes are there in the first place to interpret the blueprint.[11] There seems no way to get ribosomes—needed for translation—unless you have ribosomes—the products of translation—there in the first place.

Molecular biology has tried to ignore the fundamental circularity of genetic information processing, but the phenomenon of prions is now drawing attention to it because prions manage to reinterpret genetic information in an alternative closed loop whereby they reproduce. Prions are protein molecules which can exist in a minimum of two forms. In vertebrates, production of one form of the protein is healthy, but sustained production of the other form is associated with incurable neurological disease. In yeast and fungi, production of the alternative form of a prion-like molecule allows the organisms to change characteristics and so adapt to some environmental pressures without undergoing genetic mutation. Part of the 'blueprint' for these cells is evidently encoded in structural features of special proteins which have their own decoding mechanism. Cells carrying different forms of the yeast and fungal proteins breed true with their own distinguishable inherited characteristics, even though the cells carrying different forms of the protein are genetically indistinguishable as far as their DNA goes. This illustrates that there are continuously self-sustaining processes— subject to variation completely independent of genetic information— involved in determining biological characteristics and functions at the molecular level.

Biological stability & instability

In contrast with the implicit reductionism of molecular biology, the last ten or twenty years have seen the development of a new field of science which is generally called 'complexity theory'. It encompasses parts of theoretical biology but is broader, seeking to elucidate very general features of systems which adapt and evolve, whether they be physical, biological, social or even economic.[12] Whatever meaning is given to the term 'complexity', it generally refers to a property of systems that display 'emergent' properties (also referred to above). The emergent behaviour of the system reflects the overall pattern of interactions among the individual parts. Very simple rules of interaction between pairs of individual parts can produce complex system behaviours.

Emergent, complex behaviour only arises in systems whose dynamics are non-linear. To understand what is meant by this, we have to understand the strict scientific meaning of the term 'linear' when it is applied to a dynamic system. In linear systems, each effect is proportional to the intensity of the cause which produced it. Non-linear systems contain feedback loops, and there is no longer a simple proportional relationship between the cause and the effect. The behaviour of systems containing feedback is not as easy to calculate as that of systems whose dynamics are seen as linear. The dynamics of all biological systems are highly nonlinear. At all levels, from the biochemical interactions of a single organism to the relationships of interdependence within a whole ecosystem, the connectivity between individual biological processes is very complex and involves a host of links for inhibitory and stimulatory feedback. For example, whether one looks at the chart of chemical reactions involved in cellular metabolism or the network representing a marine foodchain, the picture is the same; each species is linked to many others that are affected when its population is increased or decreased. So, not only are emergent properties abundant in biological systems, but also emergent properties interact through nonlinear dynamic relationships to generate a multi-level hierarchy of higher order emergent phenomena, and the complexity of biological systems appears to continue increasing.[13]

In much the same way as different proteins, enzymes, etc within a single cell have functions which are related to one another—all of which hold together in a stable manner and maintain the existence of the entire cell—so, too, do the species within an ecosystem. Each species depends for its existence on the context which is formed by all the other species. The maintenance of characteristic interactions among species in an ecosystem is vital for its continued existence. The different numbers of organisms of different sorts, and the manner in which these populations change with time, cannot survive perturbations and fluctuations of all sizes, as is true for the molecular composition of cells. The particular sorts of complex adaptive systems which we find everywhere in biology seem to have some general characteristics. Ecosystems, like cells, generally have a certain robustness, resilience, and ability to cope with change. Yet, they also display great sensitivity toward some imposed changes. These two seemingly contradictory elements are ones which we see manifest, and expressed in virtually all biological phenomena.

Changes in ecosystems often appear to obey what are generally called 'scaling laws', by which biologists mean that variation in the scale of a phenomenon (for example, the size of trees in a forest) correlates in a regular fashion with the frequency or size of some accompanying effect (for example, the number of species of trees of a given size). One such scaling law applies to the occurrence of evolutionary extinctions of various sizes in the fossil record. Whatever the cause of extinctions, we find that extinctions involving only a few species are relatively common, and those involving more species are rarer. The quantitative relationship between the size and frequency of extinctions is an example of a scaling law. It can be explained in terms of a trade-off between the robustness of the ecosystem in which different species live together, and the sensitivity that such systems sometimes have towards small changes.

If we are to evaluate the ultimate effects of large-scale genetic engineering on the biosphere, we need to have some concept of how genetic change takes place in the wild, how ecosystems respond to genetic change, and how these processes are reflected in the genetic makeup of the organisms occurring within ecosystems. So far, we have not discussed genes much beyond the way they function within a single organism. We must now think about all the organisms that occur in a complete ecosystem, how they are related to one another, and how this all depends on genetic factors. We have already seen that a gene is not something which simply gives rise to a trait with all else being able to be assumed equal. Similarly, the evolutionary fitness of any one species depends on what other species are present in the ecosystem and the way in which they are changing.

Thus, the Darwinian idea of 'fitness' cannot be applied in isolation to genetic variants of one particular species unless everything else in the ecosystem is held absolutely constant—a condition which is never satisfied in the real world. Selection pressure is not something that arises simply from 'the environment' to act on single species—or single genes—individually and in isolation. Understanding the evolution of real ecosystems requires consideration of all the interacting species and the dynamic relationships between their populations. We know that an ecosystem is disturbed when a new species is introduced from a different geographical habitat, from another continent perhaps. The entire ecosystem feels the effects. The transfer of organisms across geographical boundaries during processes of human colonisation, has caused gross disruption of a number of fragile ecosystems which had existed essentially undisturbed for long periods of time, in some cases millions of years.[14] How should we then evaluate the likely consequences of transferring genes across biological boundaries?

The manner in which genes are expressed as traits in single organisms is very subtle and complicated. If we then extend ourselves to

thinking about how the genes in organisms are expressed as characteristics of entire ecosystems and their dynamics, then it is even more
difficult to find any simple relationship. The significant 'traits' of an
individual species are defined by the manner in which it interacts with
all the other species in the system, and then the interactions which make
up the characteristics of the ecosystem do not arise from the organisms'
genes through any straightforward, linear process. Thus, Darwinian
selection cannot be viewed as optimising the fitness of any single
species except within the changing context of the fitness of the other
species. Since 'fitness'—insofar as it can still be defined—reflects the
genetic makeup of the organisms that are mutually selected in an
ecosystem, individual genes in single species must to some extent
encode properties of the entire ecosystem. We are forced to conclude
that it is quite impossible to reckon what the effects of some genetic
change will be on the ecosystem that the changed organism is found in.

Consequences of genetic engineering

Genetic engineering strategies take no account of the subtle features of
biological dynamics that we have just described. Potential side-effects
of experimentation and ecological modification may be minimised to
the extent demanded by different countries' regulations governing the
release of genetically engineered organisms into the environment, but
the incremental contribution of any one release to any global process of
genetic change due to engineering is treated as if it were inconsequen-
tial. An argument can be made for ignoring broader effects if it is
assumed that you can take an organism, introduce a new gene into it,
and then assess what change has taken place by looking at the effects of
that gene on the organism alone. This approach is what is adopted in
practice—in countries where there is regulation of genetic engineering.
The genes, for example, for human proteins can be transferred to sheep
so that ewes produce the human protein in their milk with no other evident effect.[15]

However, the effects of genes are not always readily evident, as the
engineering of genetic 'knock-out' animals—like mice missing the gene
encoding PrP (see above)—demonstrates. It is difficult to detect any
difference between 'PrP knock-out mice' and normal mice. They appear
to live and function perfectly normally, and yet, because there is relatively little difference in the genes encoding PrP in different animals, it
is believed that most mutations in PrP must be deleterious and that

PrP therefore serves some vital biological function in vertebrates. Looked at the other way around, if a gene encoding a foreign protein is successfully introduced into an organism then certain cells of the organism will then contain that protein. But that knowledge is not an adequate basis for assessing all the consequences of engineering the genetic change. There may be many other subtle effects which remain hidden. When a foreign protein is introduced into a cell it can exert some influence on the way proteins are expressed or function in the cell. The most extreme form of this is the phenomenon of prions which change the manner of their own expression so dramatically that the cell is transformed to a distinguishably different type.[16] In general, the changes are not likely to be so dramatic. They may operate through cascades of interactions with other molecules in the cell, and they may only show up when the cells are in some environments; but they may nevertheless be important.

Because one of the main aims of genetic engineering projects is to produce species which can be released into the wild, or at least released for use in agriculture or other commercial activities, we need to assess comprehensively the potential ecological and evolutionary consequences of changing the genetic structure of organisms. Since the advent of Darwin's theory of natural selection, evolutionary change has been interpreted mostly in terms of adaption, the idea being that variant organisms—better adapted to their environment—arise gradually as a result of relatively slow mutation at different points in organisms' genomes. However, as the appearance of microorganisms resistant to antibiotics has demonstrated, important evolutionary changes often involve the introduction of new DNA into an organism from an outside source. It is now well established that there are many paths whereby DNA can cross from one organism to another and bring with it quite dramatic biological effects.[17] Also, because the introduction of new DNA into an organism can carry with it the possibility of major biochemical modification—as has happened in the case of antibiotic-resistant microorganisms—it is now thought that DNA transfer is a mechanism involved in the most significant evolutionary changes.

The mode and rate of gene transfer between organisms is probably one of the primary factors controlling the form of evolutionary scaling relationships. Genetic engineering has the capability of disrupting this relationship due to two effects. First, the transfers of genetic material now being accomplished are not occurring within the ecological contexts which have determined their modes of occurrence throughout the aeons of evolution. If there is any functional pattern whatsoever to the gene transfers which ecology and evolution have allowed until now, then humans are about to perturb it in ways much more dramatic than by the transfer of non-transgenic organisms between different geographically isolated habitats. The second effect of genetic engineering is that it could speed up enormously the rate of genetic change in the biosphere, collapsing the timescale defining the frequency at which changes of various magnitudes occur.

Current procedures for the legal regulation of genetic engineering, especially the release into the environment of engineered organisms, are based on assessments of changes induced in single organisms and limited investigation of the interaction between the modified organism and selected members of its proposed ecological niche. It is assumed that any unintended effects can be observed and that any unobserved effects will prove to be innocuous. The wider context is not considered, or rather it is treated as if natural genetic change were a completely random hotchpotch and its rate of no consequence.[18] The processes of adaptation, speciation, and evolutionary innovation are thought to lack any regularity, coherence or functional integrity which needs to be taken into account. That is the myth of reductionist biology. In fact, genetic engineering is a technique whereby humans are able completely to change the scale of ecological and evolutionary dynamics and destabilise the mechanism on which the pattern of structures and processes within the biosphere is based.

Genetic engineering has been seized upon as providing a convenient tool for the manipulation of nature in the service of short-term human endeavours, mostly of an overtly commercial character, whether medical, agricultural or environmental (as discussed in chapter 1). The goal of most of these endeavours is defined by the availability of the techniques of genetic engineering rather than the problem which is supposed to be solved, as has been shown in at least one case study involving the simultaneous production of proteins from transgenic animals and the generation of a pharmaceutical market for their sale.[19] In this sense it is pointless to look for 'alternatives' to genetic engineering because the technique is by-and-large irrelevant to solving problems that have not already been defined as being amenable to a genetic 'fix'. If the solution to human starvation is perceived to lie in plants whose genes confer the capability of greater grain production, then the alternative to genetic engineering—selective breeding—is likely to be slow, cumbersome, and possibly ineffective; but if human starvation can only be overcome long-term by adopting sustainable agricultural practices involving a changing diversity of plant types adapted to local conditions, then genetic engineering will be unlikely to have much of a role to play. Genetic engineering can only be used to ameliorate problematic aspects of real world situations as they are perceived by humans and biotechnological 'solutions' must be expected to generate new 'problems' for further application of biotechnology.

Humanity is faced with a choice that has more to do with collective consciousness about nature than it does with any question concerning the merits of genetic engineering.[20] Either we apply ourselves to better understanding and adapting to the dynamic processes of global ecology and evolution, or we allow the initiation of a multitude of local perturbations in the constitution of those processes which then propagate and interact, demanding an ever-increasing level of intervention in the service of changing human aspirations. With each new intervention, we will not only make a new selection of human values, but we will clumsily commit future generations to deal with the potentially disastrous consequences of that choice which, in a sense, we have by then encoded in the genes of organisms alive in the environment.

Conclusion

The processes for interpreting genetic information are so intricately complex and so diffusely distributed throughout the global ecosystem that it is completely impossible to determine in advance what the ultimate consequences of some even apparently minor genetic change are going to be. If we allow genetic engineering projects to modify the biosphere at a rapid rate and on a large scale, we can expect to perturb the dynamics of genetic change to such an extent that the changing ecological patterns which have emerged, survived, declined, and re-emerged up until our current evolutionary epoch will be completely disrupted.

Genetic engineering will not just be making some minor modification or contribution to the last three billion years of evolutionary development, but rather it will be taking the whole process and putting it into human hands—as if humans can understand it and calculate its effects so well that they can maintain whatever is needed for its well-being and functional integrity. Evolution has been judged thoughtlessly by scientists and governments to be in need of assistance so that its outcome will better suit the aspirations of twenty-first century humanity, and the tools of genetic engineering have been seized upon as the new means of improving or reconstructing nature.

To arrest this thoughtless pursuit, we urgently need a moratorium on the release of genetically engineered organisms into the environment. All releases of genetically modified organisms into the environment should cease immediately, and the ban should remain in place *indefinitely* until all the questions raised concerning the connection between ecological patterns of stability and change—and by implication, evolutionary alteration—have been given scientifically satisfactory answers, in short, have been 'understood'. We are a long way from that, and cannot as yet predict when we may arrive at such an understanding.

4

'Get out of my lab, Lois!': in search of the media gene

TIFFANY WHITE

Lois Lane sits at her desk surrounded by paper, chewing off her lipstick and peering at the clock on the wall. She's doing the science round these days, and it's taken all morning to trawl the science journals for new discoveries and research to write up under her byline. There's a press conference at Vegtech this afternoon; they're releasing a genetically engineered tomato. She rifles through her in-tray, and finds a press release about it—it'll be firmer, redder, and last longer on the shelf, they say. Lois has a lot of questions. What kind of genes did they use to improve the tomato? Aren't they inserting fish genes into tomatoes these days? Will it taste any better, or cost more? Is this the future of the food supply? Will they be labelled so consumers can choose between traditional and transgenic tomatoes? She knows that there are some critics out there who don't think these tomatoes should go on the market so soon, but they don't send press releases, and she hasn't had time to track them down.

Lois attends the conference but—like the other journalists who are present—doesn't get a chance to ask many questions. She shakes the hand of the guy in the life-size tomato suit, and accepts a free sample of the new product. She chops it into her evening salad; it doesn't taste weird, and Superman likes it. The next day she contemplates ringing those dissidents to find out what they've got against the tomato. Her editor phones and asks how many stories she's got for the next edition. She mentions the possible new angle for the tomato story. But there is a silence: 'Why bother?' he asks. 'You've got a brand new, genetically engineered tomato. It's exciting, it's novel—and why do you want to go upsetting those guys down at Vegtech anyway? We get a lot of stories from them.' She starts writing.

This Lois Lane lives in the real world of deadlines and politics. She is a different Lois from the one writing for the fictitious *Daily Planet* who would never run a rampantly positive story about genetically engineered tomatoes if there were negatives to ferret out. But the Lois of the Superman series never appeared to have deadlines. She had time to run around uncovering the truth and getting into scrapes (and clinches) with Clark Kent's alter ego. In contrast, the 'real' world of journalism leaves much to be desired.

This chapter looks at the unequal relationship between journalists and the scientific community, and the way this imbalance is reflected in mainstream science writing. It sets out to explore first, what sort of profile biotechnology enjoys in the mainstream media; second, what factors shape that profile; and, third, why that profile should matter to us.

Content & bio-images

The representative mainstream profile of biotechnology examined here was that depicted in the *Sydney Morning Herald* (*SMH*) throughout 1995. A central aim of its content analysis (which was undertaken to discover trends in the reporting of biotechnology over a full-year period) was to explore the image of biotechnology in the print media, especially in the context of a dearth of public debate.

To supplement and cross-check a substantial collection of clippings, an *SMH* database search was performed using the following search terms: 'gene', 'genes', 'genetics', 'biotechnology', 'genome', and 'genetic engineering'. The sample was not confined to straight news reports: articles were chosen for inclusion in the sample if they contained any discussion of genetic technology. This was a deliberately broad approach, designed to provide an overview of the spread of genetic issues into almost every section of the newspaper. 'Biotechnology' and 'genetic technology' were used as blanket terms for activities taking place in the field of genetics. The final yield was 118 articles.

Each article was categorised as 'positive', 'negative', or 'neutral' according to its tone. Tone, in turn, was determined on the basis of each article's overall attitude to genetic technology, and the perceived balance—or bias—of the coverage. A positive article contained such references as 'world-firsts', 'landmark studies' or 'revolutionary' techniques. It was characterised by positive quotes such as 'a nifty piece of genetic engineering',[1] or, 'experts say this is an important step towards boosting the world's food supply',[2] or, 'made using the very latest gene manipulation techniques'.[3] Quotes praising biotechnology were rarely undermined by the inclusion of negative quotes from sceptical sources. Though varied in subject and style, the positive articles were categorised as positive because they appeared to accept genetic engineering as cutting-edge technology being used for the common good, particularly in medicine. They tended not to raise questions about ethics or effectiveness in the long term.

A further division of the positive category included articles framed in terms of *breakthroughs* and *discoveries* in the field of genetics, which generally pushed the 'cutting-edge' angle. This is consistent with the view that science is often reported out of context, as a string of isolated discoveries rather than as an evolutionary process of empirical investigation.[4] In short, this style of reporting picks up only the exclamation mark at the end of a scientific sentence and ignores the explanatory material that comes before it.

An article was categorised as negative if it included critical questions about ethics and the long-term potential of genetic technology. It may have exposed the self-interested side of the industry—egos and cash—or it may have sought to provide an empathetic alternative to whiz-bang accounts of advances in the field. One father's sympathetic account of life with his Down's syndrome daughter, for example, questioned the push to eradicate such conditions, and such children, via molecular research. Negative articles did not promote unquestioning acceptance of the technology, and occasionally they encouraged outright rejection of it. Not surprisingly, these articles failed to boost the profile of genetic engineering.

To fall into the neutral category, an article must not have swung noticeably towards either the positive or negative ends of the scale. If it was announcing a genetic 'discovery', for instance, it would include quotes from the sceptics. Neutral articles might also trivialise issues, but not necessarily in a negative manner. Examples were found in the way genetic issues were treated in the 'Stay in Touch' section of the *SMH*, which often adopted a humorous approach to news reported elsewhere in the paper.

Finally all articles were further categorised according to their genetic 'story'. Twenty categories of stories were identified—from human disease to environment, to genes and ethics, and genes and cosmetic problems, to much less reported stories such as those about genes and real estate, or about genes and leisure.

Genetics reporting trends

Throughout 1995, *SMH* stories displayed a decidedly positive attitude towards genetic engineering. Sixty-seven per cent or 79 of the articles were positive, where 25 per cent depicted breakthroughs and discoveries. Only 16.1 per cent were negative, and 16.9 per cent were neutral. A breakdown of specific story categories shows a preoccupation with genetics and human disease, with 39 per cent focusing on disease conditions and medicine. Positivity resided in promises of cures, new therapies, new drugs, and genetic tests for disease conditions.

The emphasis on medical issues in the *SMH* is no surprise in the light of US research and subsequent assertions that making science information marketable to news consumers involves concentrating heavily on health. Interestingly, policy discussions of biotechnology, where directed by proponents, have relied on 'mythic' appeals to the desire for a cleaner, healthier, world.[5]

Indeed, a safe way to sell a new technology to the public is to appeal to our hopes and fears about disease conditions (even if much of the hope being peddled remains speculative). Yet despite the high focus on disease and health, few articles focused on these aspects in relation to the safety problems posed by genetic manipulation in the environment (only six stories or 5 per cent). While stories about tentative steps toward 'cures' for myriad ailments fill our papers, some of the most startling and potentially controversial advances are now taking place in the environmental sector. Some of these have the potential to adversely affect our ecosystems (as highlighted in chapters 3, 9 and 12 of this book), including those used to produce the food supply.

The only *SMH* reporting that scraped the surface of environmental issues harnessed, as a news 'peg', the escape of the rabbit calicivirus. Without that 'sensational' incident to report, it seems doubtful whether any of the background articles about other disastrous environmental releases involving exotic or foreign organisms of the past would have been written. The problems caused by insisting on a peg for science news have not gone unremarked: 'phenomena not linked to specific events, such as the greenhouse effect or the growth of an American underclass often go unreported or under-reported until an appropriate news peg arrives to supply the needed event that the coverage requires'.[6]

In addition, problematic biotechnology-environment reporting usually starts with a bang and then fizzles out, like the coverage of the US bio-hazard debate of the 1970s, which dwindled once the controversy in Congress abated: 'In the end the mass media lost interest ... there was no longer an activity to report' (see also chapter 2).[7]

After human disease, the most concentrated story category was coverage of the business and commercial applications of biotechnology. Seventeen stories (14.5 per cent) related to marketing and biotechnology deals, and they were a relatively common feature in the *SMH's* business briefs. Almost all of these stories were positive, particularly where the commercial benefits were viewed as likely to remain in Australia. Genes and human behaviour came in third at nine stories, and the personalities and historical/novelty categories were even with six stories each. The latter categories may seem unrelated to genetics, but they illustrate the pervasive nature of genetic references, which can be found in such unlikely places as the 'Good Living' and 'Spectrum' sections of the newspaper.

Interestingly, there were only four articles specifically dealing with ethics and genetic engineering. Two of those did not even reflect the *SMH* news agenda: they were letters to the editor protesting about the unethical characteristics of a feature on twin studies. The third piece reported objections raised by the Privacy Commissioner about the ethics of testing babies for HIV and genetic disorders without consent, and the fourth, buried in the *Northern Herald* (an *SMH* supplement with a small circulation serving Sydney's north shore), tackled the 'human dilemmas' posed by the 'genes revolution'. A further two features dealt with genetics and ethics more obliquely. They are 'Down to Basics' and 'Syndrome Stigmata', both first-hand accounts of life with Down's syndrome children (which Rowland also addresses in chapter 5).

The results of this study correlate strongly with Susannah Hornig Priest's analysis of biotechnology coverage in local US newspapers during 1991 and 1992.[8] Hornig Priest adopted a similar approach to dissecting biotechnology news coverage, and found (out of 600 stories) an even higher concentration of positive arguments about biotechnology (82 per cent), a higher percentage of stories focusing on economics (48 per cent), and a similarly low representation of environmental risks (7 per cent) and ethics (1 per cent). Even allowing for differences in categorisation of articles and methodology, both studies show a clear trend towards positive stories and a lack of discussion about issues that may be of concern to the public.

Front page stories

Once news is selected for publication it is subject to the rigours and vagaries of newspaper layout. Articles appearing on the front page of a newspaper have obviously been judged the top stories of the day—and as those that will sell papers. The preoccupations of a reading public, at least as they are perceived by editors, become apparent via trends in page one content.

Of all 118 *SMH* articles for 1995 on genetics technology, fifteen appeared on the front page (12.7 per cent). Of the fifteen front-page stories, fourteen were positive and one neutral. Ten of the positive stories dealt with medical issues (good news cure/therapy stories), one with world food supply, two with commercial interests, and one with sheep breeding. The neutral story may have included the comments of sceptics, but sported an appealing sci-fi headline deemed suitable for a front page position: 'Lazarus bacteria shows signs of life after 30 million years.'[9]

Public relations, the media & science

Stories generated from public announcements and deliberately accessible activities of the scientific community amounted to 51 of the 118 articles. Twenty-nine of these acknowledged either news agencies and journals used to compile the story, or indicated a straight lift from another publication. An additional twelve relied on the announced results of research programs or studies, and another ten gained their information from conferences, forums, or scientific 'events'—which the media are invited to cover via news releases.

The production demands of newspaper journalism make the use of ready-made material very tempting for the journalist. As Dunwoody comments:

> Trying to set up an interview with a scientist may be too risky: if the interview falls through, the journalist has no story. Sitting through a two hour symposium at the meeting may be similarly risky: what if the speakers drone on and nothing "newsworthy" emerges? Instead, the journalist may rely on things such as press conferences or news releases, products designed to deliver "news" quickly ... A news release will not only summarise the main point of a scientist's talk but will also deliver a few pithy quotes; a journalist in a hurry can even use the news release verbatim and no one will know.[10]

A common feature of stories reconstructed from public relations sources is the absence of interviews and challenges to claims. Some stories include quotes which are (although not openly acknowledged as such) apparently lifted from reports in science magazines or recorded during public addresses. A large percentage of the reporting of genetic issues thus relies on the scientific community's desire for publicity, solicited via media releases or through publication of research results. This also highlights the media's reliance on regular events such as congresses and annual conferences to generate news.[11]

Nelkin has charted the history and use of public relations by scientific researchers and organisations in the US to attract funding, reassure the government, and avoid unwelcome regulation (with regard to Australia, see chapter 2).[12] Using the relevant example of the emerging r-DNA technology in the 1970s, she illustrates how a strident media campaign successfully marginalised the critics. The r-DNA controversy and the measures adopted to 'kill' it are examined in more detail by Goodell,[13] who concludes that science journalism is heavily influenced and shaped by members of the scientific community and their publicity requirements.

Other work in the US has discovered a similar reliance on public relations,[14] prompting one analyst to suggest that science reporting does not represent the full range of views on a given issue because science reporters wishing to maintain access to scientist sources may adopt their sources' 'vision' of science.[15] Australia's Ian Ward wonders whether public relations effectively blocks access to news space, so that dissident voices are not only struggling against the tide of mainstream views, but are also knocked aside by the waves of publicity material rolling out of mainstream organisations with expensive PR programs. He comments:

> the widespread use of public relations may well close off the public sphere. It has made competition for limited news spaces fierce. As a consequence, many groups with limited resources and no access to professional PR have been effectively prevented from presenting their case via the media. There can be no informed rational political debate, of course, if the media—and hence the public—are given limited information organised around a narrow range of concerns.[16]

The issue of PR-generated science news is underscored in Friedman's account of the Three Mile Island accident in the US.[17] As a consultant to the Public's Right to Information Task force in the wake of the nuclear plant disaster, Friedman identified a number of factors which contributed to the local community's confusion and ignorance when things went wrong. These included the local media's mainly uncritical use of the highly technical (and thus, often inscrutable) weekly media releases from the plant's PR unit, the failure of plant PR staff to effectively inform the local media, and the deadline pressures of production in media. These factors coupled with a dearth of knowledgeable science reporters left little room for in-depth or investigative reporting. Moreover, Friedman laments the passive approach adopted by journalists writing about Three Mile Island *before* the accident occurred, saying they had enough information to detect major problems then, but didn't bother to read between the lines of the plant's regular media releases.

The influence of news values is again obvious in the Three Mile Island example. It bears some relation to the calicivirus coverage in Australia, which also escalated and became critical once there was an 'event' to justify it. Media coverage of the rabbit virus gained momentum because traditional news values applied to the story: drama, unexpectedness, conflict, personalities, and proximity. Readers did not need in-depth explanations of the science involved to follow or 'enjoy' the stories.

The convenience of PR material pitched as hard news, coupled with deadline demands, can discourage journalists from digging beneath the surface of a story or producing longer pieces suitable for the features section of the newspaper. Analysis of the 118 *SMH* articles finds that 88 stories (nearly 75 per cent) fall into the hard news or straight report category, while 26 stories are interpretive or attempt to provide background to an issue, and only four represent generalised features or opinion.

To update the previous findings, a search was made of the *SMH* archives for 1999. Applying the same search terms as in the 1995 content analysis, some 270 articles were identified—certainly up from the 1995 yield of 118. It's interesting to note (especially in light of Lois and her tomato) that 31.5 per cent or 85 of those articles are about GM foods. The debate around GM foods and their labelling featured in the *SMH* throughout 1999, and while full content analysis is required to decide how balanced media coverage was, it is clear that both negative and positive views appeared—and many feisty letters were generated on the subject. With articles about the food supply (and environmental issues as a result) up, medical stories were down to around 20 per cent—

but still mostly focusing on discoveries and 'cures'. The odd story with ethical discussion appeared, mostly generated around reproductive technology issues.

Science journos, media pressures & the news

For the average *SMH* reader, it might appear that biotechnology, especially the biomedical area, is advancing bravely into the future and shaping up as a lucrative industry, filled with therapeutic benefits for society. It would not appear to be a field fraught with ethical dilemmas, risky procedures and, if left unchecked, a field with the potential to develop in undesirable directions. Even the current ruckus about GM foods and some concerns about the environment seem to have been checked by mainly positive coverage recently on new legislation that addresses those issues (see chapters 2, 9 and 12, however, for 'negative' comment on the Gene Technology Bill 2000).

The overwhelmingly positive, or at least unquestioning, tone of most coverage points to a general acceptance of biotechnology amongst science reporters, or perhaps a reluctance to 'rock the boat' with investigative and critical stories. Much has been written about the relationship between scientists and the mass media, and such work provides clues as to why there is such passivity in science reporting. Goodell,[18] for example, notes that the press—just like the public—can be intimidated by science, so editors will require their reporters to use 'credible' (prominent and mainstream) scientific sources who are seen as more authoritative than dissidents who may lead gullible or scientifically naive reporters 'up the garden path', and undermine the credibility of a story. Science, after all, can only be right.

Can we attribute a lack of criticism in science reporting to a lack of specific knowledge on the part of science reporters? Dunwoody found that US journalists 'typically have little training in mathematics or science' which is one of the reasons that may influence their acceptance of claims by scientists.[19] An Australian study found that although some 75 per cent of 140 specialist science journalists in Australia possessed tertiary qualifications, only 12 per cent held a science degree.[20] Yet this lack of specialisation was not seen as a barrier to good science reporting, with Australian science journalists likened to an elite:

> scientists interacting with specialist science journalists should feel
> reassured that they are not generally dealing with ignorant hacks who

will shamelessly distort their research for the sake of an attention-grabbing headline, but with professional peers who should warrant their trust and co-operation.[21]

Australian science journalists who wish to remain worthy of the scientific community's time are perhaps willing to embrace the ideologies of mainstream science: chiefly, that science and technology are progressive and for the common good. It is possible to surmise then, that reports will not 'err' on the side of criticism. The high status afforded science may thus affect the balance of power between scientists and science reporters, producing a delicate relationship which has been described in terms of deference and reverence:

> The high status of science, in other words, makes criticism of it risky, particularly for non-scientists. Journalists who cover science on a regular basis are encouraged to buy into scientific truth and to pass on the versions of reality held dear by the powerful scientists of the day. Contrary behaviour is punished by a type of professional snubbing and denigration that can be costly to a science journalist, while compliance opens the doors to Nobel Prize winners. The result, of course, is that the press conveys images that are consistent with mainstream scientific views and that eschew fringe perspectives.[22]

On the whole it is difficult enough for science journalists to forge a fruitful relationship with scientist sources, let alone find the confidence to criticise what those sources have to say. This sort of imbalance cannot fail to have an effect on the tenor of day-to-day science reporting. As Dunwoody found:

> In general, science is reconstructed by the mass media as a positive force for society. You will rarely read about science's failures, about the foibles of the scientific culture. Mainstream journalism will rarely criticise mainstream science because they will be punished rather than rewarded for such behaviour.[23]

Setting the agenda

Our final consideration—the effect of science reporting on the reading public—is harder to assess. Dunwoody categorises and discusses likely media effects as follows: media stories as *signals*, and mass media as *legitimisers* and as *teachers*. Media signals are seen as the precursors to agenda-setting:

> Each day's media diet produces a set of blips on our environmental scanner. Some of the blips have little meaning to us and will vanish quickly; others may loom as topics or issues to which we may direct some of our attention. When the latter phenomenon takes place, social scientists argue that the media have "set our agenda".[24]

Ward sums up the influence of news coverage as a 'cumulative effect' which encourages people to think *about* certain issues without actually telling them *what* to think about those issues. In other words, media visibility tends to prioritise the importance of issues. An article on page one is deemed more important than an article on page sixteen; so it follows that a certain issue placed on page one carries more weight than if it was placed on page sixteen. 'The essential claim which agenda-setting theorists make is that the news media's own agenda will signal to audiences who read and watch the news that some policy issues are far more important than others.'[25]

In the case of genetics, the regularity, patterns, and tone of the coverage measured in the *SMH* content analysis would seem to indicate that genetic news, especially *good* genetic news, is high on the agenda (that is, highly newsworthy). It even creeps into the nether regions of the newspaper (for example, Good Living where traditional 'hard' news values are less important). When media signals are not quickly forgotten, they have found a place in the news agenda, and genetics certainly fits this category.

The mass media certainly do not function as an effective *teacher* where biotechnology is concerned because of the brevity of reports, the lack of alternative information, and the exclamatory approach to science stories which favours 'breakthroughs' and controversy over routine science:

> This serves to decrease informed discussions of scientific issues, because most science is not of the breakthrough or controversial variety. Instead, most science is complex and dull—two things arduously avoided by the popular press. When issues do get discussed, their framing means that lay viewers will consider each event as a breakthrough rather than as a unit in an ongoing research project. This framing plays an important role in opinion formation, and attitudes towards issues of biotechnology.[26]

This lack of informed discussion suggests the mass media plays more the role of *legitimiser* in the case of genetics because they stick to mainstream opinions. They do not adequately cover 'fringe' or other views which question the field, and thereby this can be seen to deny critics the 'social sanction' afforded positive accounts of biotechnology. Yet, while little evidence exists to support empirical claims about legitimisation in general audiences, at the basic news-gathering level it has been found that the mass media treats some views as more legitimate than others.[27] The public then consumes the 'chosen' perspectives—those considered newsworthy by journalists and editors, who are themselves influenced by their scientific sources:

> one attribute shared by the bulk of the issues that do make it onto our personal lists of "things to be concerned about" is their presence in the mass media. We—scientists included—seem remarkably willing to buy into the notion that what's in the pages of a newspaper or in the line-up of a TV news programme matters, that these few topics are somehow more important than the plethora that doesn't garner such visibility.[28]

If the science news offered is mostly the reconstructed material of mainstream scientists, we might speculate that mainstream views are certainly granted apparent legitimisation through their sheer visibility. How that visibility may influence the process by which a reader catalogues and legitimises information in his or her own mind is a question that has yet to be satisfactorily answered, and it deserves more attention. It seems obvious that the gaps in our awareness caused by the absence of ''other' views must affect our attitude to, and understanding and acceptance of, science information. We are more likely to believe a technology is safe or in our best interests if we have read little or nothing to contradict that idea.

Loge extends this argument with the idea that 'context' helps shape our understanding: 'The relative importance that the general public places on an issue is related to the relative importance that the media place on an issue. Additionally, the context in which the media place the event is likely to be the context in which the general public places the event.'[29] The Three Mile Island debacle (referred to above) points to the false sense of security that the media can unwittingly engender in a reading public by sticking to the 'safe' contextual environment of media releases where the technology is offered on a PR platter, described in rampantly positive terms with not a hint of danger. Why worry when the technology is reported in a 'safe as houses' media context with no hard questions asked?

In summary, this argument is not about media conspiracies (although the 1970s r-DNA controversy should have taught us something), but about highlighting the absence of a comprehensive debate about genetic engineering issues in the mainstream press. It is clear that scientists who talk to the media have their own agenda: legitimisation within the scientific community and amongst the public, leading to funding and support for research programs (as Hindmarsh highlights in chapter 2). Scientists' willingness to talk to journalists rarely appears to be motivated by a simple desire to inform the public for the good of society in general. It seems equally clear that the demands of production, the adherence to traditional news values, and a heavy dependence on the goodwill of scientific sources, limit the scope of science journalism and erode its potential to be of genuine use to the public.

Conclusion

Information—whether scientific or otherwise—does not come without baggage of some kind. The public needs to be aware that the science they read about in the papers is predominantly the science that someone—be it private corporations or public organisations—wants to sell, and that journalists cannot always be relied upon to seek out the whole story. As Ward suggests, 'public relations practitioners well understand that, despite the myths surrounding it, the news is not objective. It is not written from nobody's point of view, but constructed from information supplied by informants or sources'.[30]

Many of us would like to believe in the terrier-like approach to reporting we see in mythical journos like Lois Lane, but the reality is that Lois, with her dogged determination to uncover the truth, is not the kind of reporter mainstream scientists invite to poke around their labs. She is usually prevented from getting close to the test tubes, petri-dishes, and recombinant-DNA experiments which form part of the new age of genetics. Indeed, the Lois myth needs to be exploded even further, because the pressures on the modern journalist mean she does a lot less poking around than the romanticised image of Lois on the page and screen—unless you count sifting through piles of press releases.

Hornig Priest argues that it is in the interests of all parties concerned—the biotechnology community, the general scientific

community, and citizens striving for democracy in a high-tech world—
to encourage broad discussion of all the issues and view points, negative
and positive, that affect public understanding of risk. She also suggests
that the media could offer us more in this area: 'information equity in
mass media coverage of science—rather than debate dominated by
narrow institutional views—should be seen as a journalistic standard'.[31]

In the hands of policy makers, biotechnology does have the power
to affect our lives in profound ways. Unfortunately, a political understanding of related issues (or side-effects) does not always emerge in
step with new technology. At some stage in the future we will all come
into contact with some aspect of the new biology, making it likely that
there will be greater politicisation, expressed in the mass media, of the
issues involved. To some extent this was shown in the analysis of *SMH*
articles in 1999. Here, media interest about genetically engineered foods
had become obvious, even if the argument had been skewed in a pro-
GM manner. In general, the political debate at the lay level continues to
limp rather haphazardly behind rapid implementation of the technology.

Although the fearless and truth-seeking Lois would be pleased that
a more constructive debate appears to be slowly emerging for the future,
I feel sure she would be disappointed with the current lack of balanced,
open, and informed debate. With the tenacity that hooked Superman, she
would jam her spike heel in the door of the tomato lab, track down the
gagged dissidents and, armed with all sides of the story, harangue her
editor into putting the transgenic-tomato debate on the front page.

Part 2

*Eugenics,
genetic testing,
jokes & risk*

5

The quality-control of human life: masculine science & genetic engineering

ROBYN ROWLAND

Subtly, step by step, we are changing the nature of being human and eroding the control women have had over procreation. The latest male-controlled technological intervention—genetic engineering—has the potential to extend the control of masculine science over women's fertility and procreative potential in ways unimaginable a few decades ago. Camouflaging the intention to map and control human genetics with the rhetoric of 'helping the infertile', women are the experimental raw material in the masculine desire to control the creation of life—patriarchy's *living laboratories*.

The product of genetic engineering is 'man-made' [sic] and therefore 'better' than nature; and because our society does not accept the imperfect, women will be placed under more and more pressure to use all technological means offered to secure perfection. Less and less assistance will go to those who make the 'mistake' of having an imperfect child. So in the age of the perfect bioengineered and uniform product, difference—named 'defect' or 'abnormality'—will become increasingly less acceptable.

The drive for commercial and/or scientific 'success' blinds researchers and developers to the risks and moral implications of their work. Men in power, the makers of ideas and systems of control such as science, construct a make-believe world in which 'free choice' exists, in which individuals supposedly make choices about their lives unhindered by social responsibility to others. Here, the way power works is subtly hidden behind claims for personal autonomy, such as that over procreation. Belief in human control of choice is used in order to reduce human fear of risk. But risk—risk of being hurt, of death, of a handicapped child, of a sudden disability or illness—is in the nature of life. The 'control myth'—the myth that we have choice—leads people to believe that they are free, that there is no need to challenge those in power.

Mary Shelley, who wrote the book *Frankenstein* (1816), saw the perils of the control myth exercised through the use of biotechniques to intervene into natural processes. Frankenstein, the creator of a monster, lacks the imagination to envisage the implications of this desire to control the creation of life. Thinking only of his own expected glory, he muses to himself: 'a new species ... a new species would list me as its creator and source; many happy and excellent natures would owe their being to me. No father could claim the gratitude of his child so completely as I should deserve theirs'. Frankenstein's monstrous creation—who seeks only companionship and love—however, eventually destroys those loved by Frankenstein, and in the end the creator himself. In his suffering the monster cries: 'Oh earth! How often did I imprecate curses on the cause of my being! The mildness of my nature had fled, and all within me was turned to gall and bitterness'. In the popular imagination, the monster has taken on the name of his creator, Frankenstein, conflating the scientist with the monstrous. How visionary has Shelley's work been?

Human gene-splicing

To control the outcome of the child-bearing process, masculine science is now developing procedures to create 'perfect', 'unproblematic' people, through sex determination, or through the elimination of genetic illness, or through gene-enhancement of healthy normal people.

The dream of quality controlling human life raises profound questions essential to the value structure of our society. It raises ethical, social, and legal issues about eugenics and social control, and about the use of women's bodies as raw materials and as experimental laboratories. Any intervention in the structure of the embryo will ultimately affect women, because it is women who will be expected to carry genetically manipulated embryos through to term, to deliver the resulting children, and to rear them. This is carried to its extreme where it was projected recently that women may be completely displaced from the reproductive process through the making of babies outside the human body. Before the twenty-first century is over, foetuses may no longer need their mother's wombs. Instead, children may be made by the artificial creation of embryos tested for genetic diseases and subsequently grown in artificial environments to enhance their 'health and potential'.[1]

Genetic engineering can take place in the form of somatic gene therapy, when new genes are inserted into existing 'defective' cells to replace supposedly faulty genes. The resulting changes are not, however, passed on to any offspring. Germ line therapy involves changing the germ cells (eggs or sperm) or a fertilised egg, which means that changes to the genetic blueprint will be replicated in the next generation. Micro-genetic engineering is no longer a novelty. It is supposedly being developed for the purposes of 'curing' and 'preventing' genetic diseases and, as stated above, for 'enhancing' an already normally developed person. DNA 'fingerprinting' techniques are already used in the area of crime detection and immigration. Work has now been completed on the mapping of the human genome—the biologists' so-called 'Holy Grail'. The eminent US geneticist Lee M. Silver—author of *Remaking Eden: Cloning and Beyond in a Brave New World*—believes that the Human Genome Project is 'probably the most important thing that has happened to humankind'. He believes it will allow us to soon know how all the 100 000 human genes make us work and operate and who we are.[2]

Molecular biologists and geneticists are developing gene probes to determine the genetic bases of some diseases, such as sickle-cell anemia and thalassaemia, Duchenne's muscular dystrophy, Huntington's disease, cystic fibrosis, Lesch Nyhan disease, Down's syndrome, manic depression, and cancer. In 1990, the first trials for human gene therapy approaches were approved.[3] Seven years later, more than 200 clinical trials were underway worldwide—with hundreds of patients enrolled. Although gene therapies are predicted to become in the not-too-far distant future as routine as heart transplants, so far there has been no successful outcome. Apart from their claims that they will find genes which will cure diseases, scientists believe that they will be able to prevent certain health problems from appearing. For example, they are discussing an early-warning system for preventing diabetes, as well as fighting viral diseases and stopping neurodegenerative diseases.

It is disturbing that the research is moving from a discussion of curing genetic diseases to preventing all manner of problems which might arise in children such as asthma or stress. Heart disease and cancer—which have known environmental factors involved—are also considered as genetic problems. Increasingly, health and behavioural problems are linked to a genetic cause. For example, predisposition to problems in the menopause is now discussed as genetically predetermined. This attitude may lead to invasive medical procedures for problems which may not even be biological, let alone genetic. Commercial interests are developing 'kits' for the genetic screening of a person's genetic code for possibly defective genes. It is clear that when translating genetics into a testing procedure, scientists are reducing problems to a simple analysis, looking for single-gene causation of possible multiple-gene defects and ignoring environmental or psycho-somatic factors.[4]

Once a supposed genetic problem is detected, the question is, what action needs to be taken to restore the person to health? To date, discussions have concentrated on somatic therapy. But some scientists argue that it should be replaced by germ line therapy, since its effects are passed on, whereas somatic gene therapy will only deal with the problems for the immediate individual. Germ line therapy, some claim, will deal with 'cleaning up' the human gene pool in general and will therefore be more 'cost-effective'. As one commentator put it: 'Gene therapy could be used to alter the genetic blueprint of whole populations'.[5]

A more recent development along such lines, since the announcement of the birth of Dolly the cloned sheep in February 1997, is quite possibly the gene-enhancement of cloned humans in the not-too-distant future. As Silver commented in 1999:

> We're beginning to figure out what genes are involved in causing differences in personality and behaviour in animals, and what this means is that by figuring out what these genes do, in a few years we'll be able to get a sense of what are the genes involved in behaviours and personalities of humans as well. And that has spectacular implications because if we can find genes that might be involved in predisposing people to be more or less aggressive, or people to be shy, or people to be novelty seeking or more curious or less curious...(then) these are all traits that parents might want to modulate to give more or less of to their children ... [6]

Silver indicated that the idea was to start off with 'designing' babies who would never be prone to heart disease, obesity, or diabetics, and then to slowly move 'down the line' to develop babies, for example, who would not be prone to alcoholism. He would then move to modulate personality traits such as shyness. Shyness—apparently, according to Silver—causes unhappiness.[7] Is homogeneous happiness the future?

Despite the common message from the biotechnology experts that modern biotechnology is user-friendly, the Feminist International Network of Resistance to Reproductive and Genetic Engineering (FIN-RRAGE) argues that gene technologies are not 'error-friendly'; that is, if errors are made there is no way that the technology can be reversed. With somatic gene therapy, the possibility exists of an increased multi-plication of cells which was not desired, and the development of cancer. Once inserted, the genes are uncontrollable and may embed themselves in the wrong place. The inserted gene may be inappropriately expressed, or may affect other genes around it. Retroviruses which are used to carry the DNA in gene splicing may also be dangerous.[8]

Perhaps a more fundamental problem with the work in genetic engi-neering is that it ties behaviour back to genetics. For many decades now, feminists, and many social and holistic scientists, have been working to encourage a move away from biological determinism (or reductionism as Wills and Love both discuss in this volume), arguing that masculinity and femininity are not biologically determined, that dark-skinned races are not less intelligent than white, and that the poor are not poor because they somehow biologically deserve it. Science has shown the public that heart disease and cancer are related to stress and to diet. What happens to these variables in the great genetic push? Stress should not be dealt with by changing the genetic structure of a human being, but by changing the environment which produces stress and changing the ways that people learn to cope with stress levels. As Jacques Testart—a leading specialist in the test-tube baby business in France—says of geneticist's pronouncements about identifying and replacing genes through projects like the Human Genome Project and gene therapy, 'You can't label a gene as being good or bad. A gene is part of the complexity of life and saying it's as simple as a mecanno set is fraudulent'.[9]

Dangers of genetic engineering

Health risks have occurred for those working within genetics. Molecular biologists at the Pasteur Institute in Paris developed various cancers while working on techniques of genetic engineering. Two died. In a fur-ther incident, farm workers in Argentina were infected with a genetically manipulated rabies vaccine which was being tested by the Americans without the permission of the Argentinan government.[10]

The approach of scientists to environmental hazards and pollution is increasingly to determine a method by which those likely to become ill through environmental contamination can be screened out of certain workplaces. What seems to be the ultimate insanity of a science created in capitalism is the argument that we do not need to clean up the environment, instead we make people genetically more able to deal with pollution. Scientists have studied genetic predispositions for sensitivity to drugs, pesticides, and heavy metals, as well as studying the sorts of people who may have an increased risk of developing cancer, heart disease, and allergies (as Gesche also comments upon in chapter 6).

But who will control the products of this genetic engineering and its application? Scientists themselves will not have ultimate control. Upon commercialisation, control lies in the hands of the state and multinational corporations. The very fact of patenting animal, plant, and human genetic material indicates that its purpose is to generate profit. In 1997, Melbourne pharmaceutical company AMRAD was reported to have patented a mildly retarded child's DNA sequence, after researchers had discovered an unusual genetic kink on a chromosome.[11] This 'centromere' could be cloned and used to build artificial chromosomes (gene carriers) to deliver gene therapy 'to potentially alleviate suffering from such inherited disorders as haemophilia and cystic fibrosis'. The patent came under instant criticism from Australian GeneEthics Network director Bob Phelps, who said 'that the rights of an anonymous donor had been extinguished and that drug companies stood to make "hundreds of millions of dollars"'. Phelps added that 'there should be a ban on the patenting of human genes because they are "discovered, not invented"'.[12]

As genetic testing becomes more effective to predict potential health hazards for people, many companies are intending to develop vaccines and gene therapies to ameliorate or prevent these conditions. As a geneticist at the University of Utah is quoted as saying:

> There are two levels: first, you test with markers to define predispositions, and having implicated [certain] genes ... you study how these genes work. Then new means of invention will suggest themselves, and you market the drugs you can take to mitigate the predispositions.[13]

Those interested in using predicability tests include various multinational and insurance companies. A study on genetic screening in 1992 by the London-based Nuffield Foundation argued that 'insurers should not be free to cancel existing [health insurance] contracts in the light of new knowledge, and that liberal societies should acknowledge an obligation towards those for whom commercial insurance is declined'.[14] Today, some of the biggest US companies are using genetic tests of workers' blood and urine to reveal so-called 'susceptibility to harm'. The screening of workers for genetic imperfections shifts the responsibility for workplace illnesses to the worker rather than to those controlling the work environment.[15] In a 1993 *Time* telephone poll of 500 adult Americans, some 87 per cent disagreed with the proposition that it 'should be legal for employers to use genetic tests in deciding whom to hire'.[16]

Testing can further be discriminatory where medical schools might exclude older students because training is costly if the student is 'likely' to have a heart attack and die young. Similarly, airline pilots are another group that could be singled out. Dr Kenneth Pagan from California indicated that though this kind of screening is possible, the opposite could also be true. 'In factory jobs for which training was cheap, but pensions were costly, such jobs could be relegated to people likely to die young'. As early as the 1970s, companies began to deny insurance to people who were carrying sickle-cell anemia and yet were healthy.[17] In October 2000, Great Britain became the first country to allow insurers to use genetic tests to identify people with Huntington's disease. Australian insurance companies have put forward a similar proposal.

There is a possibility with DNA fingerprinting and with the work on mapping the human genome that in future, people may be given 'gene credentials' indicating that the person is 'predicted' to develop depression or stress-related disease, or has an inherited disease. When prominent Australian science reporter Robyn Williams interviewed the Human Genome's director he was told, 'yeah, it is a possibility ... they will know an awful lot about people which certainly will have the propensity to invade people's privacy. My feeling about it is that the net outcome will be much more good than harm, but there is a risk of harm ... You have to make your own choice'.[18]

Predictive genetics could also be used during divorce proceedings to ensure that child custody is not given to the spouse predicted to develop disease. Life insurance companies could refuse to allow insurance for those who indicated such predispositions, or could have a cutoff point at which a person was expected to die. All manner of abuse lies open if this technology becomes more widely available and accepted. Much of this work will be done under the guise of 'genetic counselling' set up altruistically to assist people to make decisions about whether or not they should bear children—and increasingly which kind of child they *should* bear. The Law Reform Commission of Victoria summed up:

> The development of genetic counselling and therapy may raise other problems for patients. Pre-natal genetic testing is likely to become more common, perhaps even routine. Media publicity of birth defects and the new technology available for detecting and treating them will increase patients' demands for pre-natal testing. There will be commercial pressure from the biotechnology companies producing testing materials. Physicians may urge that pre-natal tests be undertaken to reduce the chance of negligence suits. New sub-specialities in medicine, such as genetic counselling and clinical and laboratory genetics are developing to serve the new market. Government financing of public health programmes may focus on prenatal genetic health issues. Medical insurance benefits may be conditional upon pre-natal screening.[19]

Some, however, argue that genetic engineering and DNA finger-printing will greatly advantage society. McLaren graphically presented the 'pro' case when she wrote about the genetic problems of some children:

> some may expect to live for two or three years (as in Tay Sachs disease), for ten or twelve years (Lesch Nyhan) or for twenty years or more but with rapidly increasing disability (as in Duchenne's muscular dystrophy). Some conditions cripple either the body or the brain (Down's syndrome, for instance) but do not kill. Others do not appear until middle age (Huntington's disease), so people with an affected parent may live all their lives in dread of developing the same disease.[20]

The prospect of invasive genetic work on people who have genetic diseases raises the issue of our attitudes to disability in general. Special-needs people are seen either as hopeless beings with no quality of life, or as stoically carrying on against all odds. But these people themselves point out that there are a many strengths, weaknesses, and capacities in the so-called 'disabled person'; society often stereotypes the problem.

Alison Davis, born with spina bifida, argues that she is not less of a human being because her legs do not work. She argues that discussing a disability in isolation from the individual who has it encourages people not to understand disability but to abort the problem. Davis argued that one of the reasons that abortion after pre-natal screening is advocated is because it is more 'cost-effective than caring'.[21]

Marsha Saxton, who also has spina bifida, has argued for the rights of disabled people. She argues that the approximately 40 million such people in the US have been 'silenced'. She challenges the following assumptions: that having a disabled child is wholly undesirable; that the quality of life for people with disabilities is less than that for others; and that we have the means to humanly decide whether some are better off never being born.[22]

She points out that most people have not had contact with special-needs people, because of the latter's institutionalisation; this distance has created false stereotypes and negative images. Her advice to a woman considering pre-natal screening with the intention of aborting an 'abnormal' fetus is to ask herself whether she has sufficient knowledge about the specific disability itself, whether she personally knows any disabled adults or children, and whether she is aware of the distorted picture presented to people of the lives which disabled people lead. The major issue involved, she argues, is the availability of societal resources for the parents, the family, and the community to assist the child to develop to its fullest potential. She writes, 'we will most likely never achieve the means to eliminate disability: our compelling and more profound challenge is to eliminate oppression'.[23]

Beck concurs with the view that gene therapy is not just a simple way of removing congenital diseases.[24] He argues that it is, in fact, an attempt to 'abolish the congenitally diseased' and to reduce people's lives to a single characteristic. Future generations are being determined not by dealing with a human body but with 'what seems to be dead matter that can be arbitrarily and "painlessly" selected and instrumentalised according to certain features'.

But the varying degrees of disability brought about by different genetic problems, cannot be assessed in any way by pre-natal screening. Discussing two of her children who were born with cystic fibrosis, Jackie Andrews indicated many affected children have happy lives and live into adulthood. Her daughter Melanie died at the age of ten after a year's serious illness and mild disability throughout her life. Her second daughter Alex at the age of eleven had just won the medal for her class in judo. Given the opportunity, Jackie would not have opted for abortion of either of her daughters. She says that 'in spite of her short life, Melanie enjoyed her 10 years ... and the unborn child, like Alex, may have something to achieve'.[25] The life of the paralysed Christopher Nolan is another example of the achievements possible when loving care and assistance are available. Nolan has produced two widely acclaimed books: his second (1987) won the Whitbread Book of the Year award in England.

Further rebutting the biotechnological vision of the desirability of gene therapy is that certain genetic illnesses can actually produce positive effects, as O'Neill reported:

> Those who develop sickle-cell anaemia have two copies of the defective in their chromosomes. But for those who inherit only one copy of the sickle-cell gene, it confers a better ability to survive malaria—a disease common in those parts of Africa where sickle-cell disease also is endemic.[26]

In her study of women who have used or refused amniocentesis—diagnosis for chromosomal abnormality in the fetus—Katz Rothman cited women who argued for acceptance of the risks in having a child, rather than screening them. Summing up the dilemma she wrote: 'For some people, this is the answer: we accept our children, we comfort them, we do what we can, but we cannot take their pain and their lives for them. For others, that is not enough. For some, it is better not to have a child than to have a child that will suffer.'[27] But the hideous irony is that no woman's decision can ensure this anyway. The screening technology is not failsafe, and cannot in any case guard against disability which is not caused by genetics. In many countries, the main predictors of disability and disease in newborn and young children are poverty and the extreme youth of the mother.

Eugenics

The desire for human perfectibility leads inevitably to selective breeding. As Testart said over a decade ago, 'With genetic progress, the way is open for eugenics'.[28] Ten years later, in 1996—with rapid advances in molecular genetics—Wilkie was able to project a disturbing future eugenics scenario with the looming spectre of designer babies:

> Sharon had started asking The Question: Where did I come from Mummy? As she patted her daughter's blonde hair (catalogue number: HC 205) and looked into her perfect cornflower-blue eyes (catalogue number: EC 317), Tracy decided that, in the year 2095,

there was nothing to be squeamish about ... Trevor and Tracy ... had gone to the Ideal Baby exhibition ... combed through months of back issues of *Genes and Babies* magazine ... window-shopped at the Swedish genetic superstore ... What more could loving parents do for their child? Although, she reflected, it might be better not to mention ... that Trevor and Tracy had not been able to afford the most expensive high-intelligence genetic profile and had to settle for a cheaper model (catalogue number: IQ 200). Sharon would therefore always be less intelligent than Charlotte next door, whose grandparents had taken out a second mortgage to help purchase for her the genes guaranteeing ultra-high-intelligence (catalogue number: IQ 300).[29]

Eugenics was a term coined by Francis Galton (cousin to Charles Darwin) in 1883. Eugenics is derived from the Greek term *eu* which means 'good' and *genic* from the word for 'birth'. The eugenics movement was very strong in England and North America as well as in Germany and Sweden. In the US in 1910 a Eugenics Record office was established and had two outcomes: compulsory sterilisation laws and the restriction of immigration. By 1931, thirty States had such laws. Compulsory sterilisation applied to 'so-called sexual perverts, drug fiends, drunkards, epileptics and "other diseased and degenerate persons"'.[30] By 1935, 20 000 people in the US had been forcibly sterilised. Sterilisation laws remain on the books in twenty States. With the first institute for racial biology established in Sweden in 1922, sterilisations on the basis of eugenics goals peaked in 1944 with 1437 operations. Recent revelations indicate that Sweden continued its sterilisation program until 1976, by which time social and economic grounds had replaced eugenics grounds for the operations.[31]

In Australia, eugenics gained many adherents also before the development of the policies of Nazi Germany. It particularly appealed to those who believed that poverty and misfortune were the result of immutable weakness of character. An endowment for mothers was rejected by many on the grounds that it would encourage people who were not desirable parents to breed. 'What the world wants is the sound teaching of Eugenics; healthy parenthood ... and not the reckless propagating of a diseased and undesirable race'. Unfortunately, some feminists were involved in this philosophy. Millicent Preston Stanley in her speech to the New South Wales Legislative Assembly in 1927 advocated legislation to segregate 'mentally defective persons, for the purpose of effecting an improvement in the race-stock'. Statements by the Mothers Club of Victoria followed this line, and Mrs Priestley was quoted in the *Argus* on 10 September 1935 as saying, 'It had been estimated that in one generation mental deficiency would be reduced by half if sterilisation was legalised. In Australia there was power to make the race we wished. It should not only be a white race, but a race of the best whites'. A movement supported by well-known feminist Jessie Street encouraged the compulsory exchange of health certificates between people who were going to marry. The Racial Hygiene Association issued posters indicating that preparation for marriage without physical tests at a pre-marital clinic was a sign of blindness. These tests included blood tests, chest x-rays, tests for diabetes, and tests for inherited diseases. Women were reminded 'that motherhood was "not an instinct but a science"'.[32]

In Nazi Germany, the philosophy extended from the sterilisation of 'mentally deficient' people to their literal extermination, and then on to the extermination of other so-called 'undesirable' groups, including gypsies, homosexuals, and Jews. Doctors involved in this extermination, which Lifton described as 'killing as a therapeutic imperative', managed to reconcile their work with that of their Hippocratic oath as doctors. Astonishingly, one Nazi doctor said: 'Of course I am a doctor and I want to preserve life. And out of respect for human life, I would remove a gangrenous appendix from a diseased body. The Jew is the gangrenous appendix in the body of mankind'. But Lifton is not just talking about doctors through the Nazi period. He writes:

> One need only look at the role of Soviet psychiatrists in diagnosing dissenters as mentally ill and incarcerating them in mental hospitals; of doctors in Chile serving as torturers; of Japanese doctors performing medical experiments and vivisection on prisoners during the Second World War; of white South African doctors falsifying medical reports of blacks tortured or killed in prison; of American physicians and psychologists employed by the Central Intelligence Agency in the recent past for unethical medical and psychological experiments involving drugs and mind manipulation ... doctors in general, it would seem, can all too readily take part in the efforts of fanatical, demagogic, or surreptitious groups to control matters of thought and feeling and of living and dying.[33]

The emergence of genetic engineering has caused similar attitudes to again become visible. A former health systems analyst in the office of the former American Surgeon-General suggested that the existence of mentally deficient people prevents the solution of social problems. He said: 'If we allow our genetic problems to get out of hand, we as a society run the risk of over-committing ourselves to the care and maintenance of a large population of mentally deficient persons at the expense of other social problems.'[34]

Is there, however, a need for enforceable legislation? A survey of consultant obstetricians in England showed that 75 per cent of them insisted that women agree to abort any abnormal fetus they might be carrying before an amniocentesis was carried out.[35] Against this background, how tempting would it be for parents to also abort otherwise healthy offspring who do not fit into a fashionable idea of the perfect child? As Hubbard has pointed out, 'In this liberal and individualistic society, there may be no need for eugenic legislation. Physicians and scientists need merely provide the techniques that make individual women, and parents, responsible for implementing the society's prejudices, so to speak, by choice'.[36]

Whichever way it is organised, through legislation or 'choice', the outcome of eugenicist attitudes means selecting humans of value and non-value. Who sets the criteria? And what of the role of women, who ultimately will have to carry genetically manipulated embryos to term or who will be given the 'choice' of pre-diagnosing the health of their children? Wheale and McNally write:

> The gap between the Daedalean power of this revolutionary new science and technology and our inability to foresee its consequences creates a moral duty which can only be executed when the utopian ideal of perfectibility, which is embedded in the scientific endeavour, is superseded by one of greater responsibility and accountability.[37]

Science and commerce have failed to be either responsible or accountable to women. Instead, a gradual softening-up process continues to move the public towards accepting genetic engineering.

Resistance

For those who want to resist the Bio-utopian world that patriarchal science is eagerly leading us to, what forms of action are open to challenge, impede, or stop it? Are there avenues by which women (as well as men) can reject the equation of personhood with fertility, especially under the impact of genetic engineering and the masculine dream of quality control. Resistance especially lies in self-empowerment of women at both the individual level and at the social level. A key area is raising the voice of women and of the community—people need to become more articulate on the issues involved. Information about the issues needs to become more accessible so that we all understand better what science is attempting.

FINRRAGE, created in 1984, with members currently in 40 countries, has developed four main strategies for creating change in this area: the open exchange of information and the development of activism; the development of research into this field to create feminist theory and practice; the publication of information and research to increase public awareness and debate; and, pressure through publicity and lobbying to influence legislation and public debate.

Although scientists profess that they are seeking public debate, when the community creates or seeks legislation to control their work they actively work against it in order to reassert scientific freedom (as Hindmarsh highlights in chapter 2). Scientists feel they are a law unto themselves. They claim the ability to self-regulate, yet the evidence of their actions often does not support their rhetoric about the benefits of reproductive and genetic engineering. Clearly, where there is so much profit to be made, biomedical researchers cannot be the guardians of ethical values. Unless the community takes control through legislation and active participation in decision-making processes, medical research in association with commercial interests will continue to determine the nature of our society.

Today, there is an opportunity to stop the movement which is occurring towards the dehumanisation and commodification of life. It is possible to stop the processes which are turning women into living laboratories and children into 'perfectible' products for sale or exchange. Many share the desire for a more humane vision of the future. A unification of that purpose, a belief in the possibility for change, and a concern for future generations is a solid base from which to counter the thoughtless pursuit of scientific control and the greed for profit, which reproductive and genetic engineering represent.

6

Genetic testing: a threat to privacy

ASTRID GESCHE

Jackie and Emma are sisters. They recently took part in a research program looking for alterations in the gene BRCA1, that confers a susceptibility to breast and ovarian cancer. They learned that they both carried the altered form of the gene, which put them at an increased risk to develop the cancers. Subsequently a mammogram revealed that Emma had a small cancerous lesion in her breast, while no cancer was detected in Jackie. Both sisters decided to undergo a bilateral mastectomy in the hope that this would reduce the risk for breast cancer. Jackie and Emma felt strongly that they benefited from knowing this genetic information. However, they were fearful that their genetic status could be used against them and their family by insurance companies and employers.[1]

The science that studies genes, how they work, how they are passed on from parent to offspring, and what effect they have on an organism is called genetics. Genetic factors are said by many scientists to guide most human characteristics, from our outward features and behavioural attributes to our susceptibility to disease. Traits such as eye-colour and height, personality and intelligence, and health and disease are all influenced by our genes. Genetic differences can have a variety of implications for individuals and for society. The story of Jackie and Emma above is one example—it is a story of our time and of our future.

The ability to study diseases at the gene level is enabling us to understand their origin and development to an extent not possible a few years ago. Today, we know of some 6000 conditions that are partly, or entirely, the result of abnormal or missing genes, or abnormal chromosomes. For the majority of genetic diseases, there is currently no treatment and no cure. Some people, like Jackie and Emma, take radical measures to reduce the risk (however there is no guarantee that it is reduced through such measures). The sisters based their decision on the nature of their mutation. Sometimes, however, simple life-style changes suffice to modulate the effects of genetic aberrations or abnormalities.

Over the last few years—and due to the multi-billion dollar international 'mapping' project of the Human Genome Project (HGP)—we have found out a great deal about genes. The HGP started in 1990. It set out to identify all the human genes in a genome and to construct a detailed map of their three million DNA base-pairs located on our chromosomes. A 'working draft' sequence was announced on 26 June 2000. Already we have gained astonishing insights into the 'private lives' of genes. The emphasis is now rapidly shifting from gene structure and gene sequence to gene function.

Even though geneticists have learnt a lot about the molecular origin of a particular disease, they often have no clear idea of the extent to which genes are responsible for disease symptoms or of how to cure genetic abnormalities. Therefore, many geneticists, scientists, physicians, ethicists, and others ask whether people should be tested for particular genetic predispositions at all. Do individuals really benefit from the information? Genetic information can cause anxiety, stress, and feelings of guilt. Could the information cause more harm than good? Could unfavourable genetic information also lead to discrimination and stigmatisation? Jackie and Emma were fearful that information about their mutation, if not kept secret, could have serious consequences for their professional and private lives. The sisters' story points to unsolved ethical dilemmas: how to keep our genetic make-up and predispositions private and away from prying eyes; and how to ensure that others do not misinterpret and misuse the genetic script of nature and cause us and our families untold harm.

This chapter summarises some of the problems associated with privacy and genetic information in the medical context (others are spelt out by Rowland in the previous chapter). Essentially, there are many good reasons why legislation is needed to protect our genetic privacy. Legislation should protect medical records; address the sharing of genetic information and the consent process; prohibit the collection or analysis of genetic samples without consent; and guide and protect the collection, storage, access, and destruction of genetic information.

The field of genetics (and, especially, molecular genetics) is developing rapidly, with new developments being reported regularly by the media (as White elaborates upon in chapter 4). New techniques are being discovered, and applications made, in a bewildering array of settings. The following is a signpost to a genetics future still open to many possibilities.

The special place of genetics in a person's medical history

Family members share a common gene pool. At times the inheritance of traits can bring joy, but occasionally they bring sorrow. As researchers probe deeper and deeper into the world of genes, they discover that many more diseases than might have been expected appear to have a genetic component. Jackie and Emma's radical actions were one consequence of this expanding knowledge. Their life-threatening predisposition to breast and ovarian cancer could only have been exposed so early with molecular technologies. Modern genetics may have spared them an early death. But why the fear? What is so special about genetic information?

Several unique aspects of genetic disease distinguish it from other types of diseases. First, there is the special mode of transmission and the permanency of a mutation. Genetic diseases are transmitted from parent to offspring. They are not passed on from any unrelated individual to another—as in the case of infectious diseases. Genetic aberrations, if expressed, do not respond to treatment as do (most) contagious diseases. Mutations remain with the affected person for life.

Then there is the *predictive* power with regard to relatives. For families in which at least one person is affected by a genetic disease, the chance of recurrence in other members can often be predicted by statistical probability studies. In the case of Huntington's disease, a person born into a family with a known history of the condition has a 50 per cent chance of carrying the defect. Huntington's disease is a progressive, debilitating, disease of the central nervous system. It is characterised by involuntary movements, loss of motor (voluntary) control, and dementia.

Another genetic mutation—cystic fibrosis—affects about one in 2500 Australians. Among the classical features of cystic fibrosis are chronic abnormal secretions of a sticky mucus that blocks the lungs, pancreas, and other parts of the body, and often causes an infection of the lungs. The genetic defect in cystic fibrosis lies in a gene called *CFTR*. The defect is unnoticeable in a carrier. In carriers only one of the two inherited copies of the gene is defective. The second healthy copy masks the defect, and such individuals do not suffer from the condition themselves (and, as Rowland points out in chapter 5, this can be positive in another way). The defective copy can, however, be passed on to their offspring. Thus, if two carriers produce children, each child has a one-in-four chance of being afflicted with the disease.

In the case of many genetic mutations, we cannot determine when and whether the disease will develop. Unlike Huntington's disease or cystic fibrosis, some abnormalities might not manifest themselves until additional elements come into play. In some patients with a predisposition to lung cancer, abstinence from tobacco and other air pollutants might reduce or totally alleviate the risk. In arteriosclerosis (a chronic disease with hardening of the arteries which hinders blood circulation), where more than one gene is implicated in the development of physical symptoms, many non-genetic factors—such as diet, exercise, stress, and smoke and alcohol intake—influence the outcome of the condition. This means that a person carrying a genetic abnormality can only be said to have a propensity to develop a condition, but other factors—such as environmental influences or other gene sequences—can modify, or even eliminate a negative outcome.

Lastly, there is the predictive power of genetic information with regard to a person's health. The descriptions above of Huntington's disease, cystic fibrosis, and breast cancer illustrate that the predictive power of genetic testing is substantial, and that the test result can concern others apart from the individual tested. It will have a profound effect on the way we view ourselves and how others respond to our family and us. But the power of predictive testing should also concern the (apparently) healthy. The Human Genome Project is indicating that many common illnesses have a genetic component. They are the so-called polygenic diseases that range from cancer to cardiovascular disease, from asthma to psychiatric illnesses. Since genetic alterations also happen to be the most common causes of illness in the Western world, it is understandable why more and more people are urging scientists and policy makers to protect the privacy of genetic information.

Genetic registers, DNA databanks & legislative protection

In the medical sector, computerisation of medical records and networking of computers is widespread and increasing. We are now seeing the first medical information services on the Internet. Are the records safe? Hackers have already penetrated hospital and research computers; intercepted the transmission of data; changed patient data; accessed confidential computer files that contain sensitive information about patients and their disease; and implanted computer viruses and deleted valuable research records.[2] Their actions pose dire—at times life-threatening—consequences for patients. Criminal tampering by hackers constitutes a serious breach of the code of ethics on privacy and confidentiality that exists between patient and medical professional.

Many countries—among them Canada, Germany, France, and Great Britain—have data protection agencies that provide safeguards against misuse of information held in government agencies. Australia, too, belongs in this category. Data collected and stored in Australian government agencies are protected by government legislation. Yet, Australia is still waiting for coherent privacy laws to cover computerised data collection and storage in the private and state public sector.[3] What makes matters worse is the fact that common law in Australia does not recognise violation of privacy as a tort (that is, as a civil wrong or injustice).[4] This makes the patient very vulnerable and very reliant on the existing, but unenforceable, codes of ethics associated with various groups of health professionals and para-professionals.

The confidentiality of patient information is the foundation of trust in any doctor–patient relationship. Sometimes it is necessary for a doctor to pass on personal patient information to other health care professionals who are involved in treating the patient. If a patient does not consent to the passing on of information, the doctor usually respects the patient's wishes after due counselling and after explaining the likely consequences of such a refusal. This may be acceptable as long as the patient is informed and competent to make a choice, knowing full well the possible adverse consequences to himself or herself alone. However, when the health and future of other genetic relatives are also implicated, nondisclosure becomes problematic. The nature of genetic information is such that it comprises information not only about a patient's present and future physical and psychological health, but also points to the genetic status and future health of members of the wider family.

For a doctor, respecting the privacy of information about a patient is voluntary. The Australian Medical Association's (AMA) Code of Ethics—despite its revision in 1995—does not impose a legal obligation on the doctor to adhere to that code. No statutory restrictions exist on the exchange of sensitive medical information from doctor to doctor, although the code does insist that patient consent must be obtained beforehand.

Medical records are not the only place where genetic information can be found—it is also held in genetic registers and in DNA databanks. In the clinical and research context, genetic registers and databanks store health information about families with certain genetic illnesses. They can also store DNA samples. The primary function of these registers is to collect epidemiological data about certain genetic conditions in order to assist in the diagnosis, counselling, and treatment of at-risk individuals. They also collect, register, and store samples which can provide researchers with biological material for their research.

Most medical research projects, including genetic research, are conducted under the umbrella of the National Health and Medical Research Council (NHMRC). In its *Aspects of Privacy in Medical Research* published in 1995, the NHMRC requires strict adherence to its framework for handling private information. The overall premise is that confidentiality is of utmost importance—that data are not to be released to third parties without consent. If they are to be released, they can only be used for the purpose for which they were collected. The NHMRC further asks the ethics committees of individual institutions to scrutinise and oversee a project from its design stage onwards. The aim of the ethics committees is to try to strike a balance between the privacy rights of the individual and the health interests of the public. The guidelines however cannot be enforced in law. As long as monitoring remains based on the goodwill of all parties, any infringements of the privacy of individuals cannot be fully pursued legally.

If genetic registers and DNA databanks could be protected from abuse, and if they could be managed in such a way that the individual and the family donating the material would remain the absolute custodians of that information, these depositories could well be an invaluable resource for counselling, treatment, and research. At present, Australia is still attempting to achieve that goal. A Canadian study in 1995 revealed problems and substantive deficiencies arising from the lack of formal and uniform DNA-banking policies.[5] The study found that most DNA banks lacked written internal policies and written agreements on how to deposit genetic information. This, the authors argued, might lead to future misunderstandings with depositors or to unforeseen legal liabilities.

One consequence of the expanding knowledge derived from the Human Genome Project is that the demand for genetic information will increase in the future. Already researchers look to other potential resources that could supplement existing DNA databanks. By far the largest such resource could be the tiny samples of blood taken from newborns, dried, and then used in Guthrie tests originally used to screen newborn babies for phenylketonuria—a genetic disease— and extended today to screening for disorders such as cystic fibrosis, galactosaemia, and hypothyroidism.[6] These samples constitute a vast repository of genetic information with an enormous re-testing potential (as dehydrated DNA is very stable for a long time). For example, in Australia, the Royal Brisbane Hospital stores Guthrie cards dating back to 1986. In other laboratories the samples stored are much older. Eleven respondents in a 1993 American survey of hospitals reported that they had been saving their Guthrie cards since the 1960s. Seven laboratories had amassed in excess of 500 000, with four of these reporting collections of between one and five million cards, and one, six million cards.[7] Such storage has generated additional concern among some people with regard to the privacy of medical records.

Considering the growing interest of stored Guthrie cards, written, legally binding regulations are needed urgently. They need to accord stringent privacy protection and assurances as to the proper management of these registers, including written regulations on third-party access to the retained Guthrie cards. As long as insurance companies vie for access to genetic information, the public has to be on its guard. For example, should Guthrie samples ever be reclassified as 'genetic test' material, the assurance by insurance companies that they want the results of only prior genetic testing for new applicants (rather than the routine screening of all new applicants) would take on a very different meaning! Everyone born from the mid-1960s onwards would fall into this category of prior genetic testing.

Genetic privacy & insurance

Life insurance companies everywhere have been quick to realise that genetic testing for disease susceptibility could be to their advantage by providing an additional criterion in their assessment of mortality risk.

The medical and life insurance system in the US is largely built around private insurance companies, many operating in close association with employers. Thus, in the US, the potential for discrimination in health and life-insurance coverage on the genetic basis of common disorders can have devastating consequences for affected individuals (as also highlighted by Rowland in this volume). One early—and well-publicised—case of discrimination in America has been that of Theresa Morelli, who applied for disability income protection with a large insurance company. Theresa was a young, healthy law student of the University of Akron in 1989. Her father had been diagnosed earlier with Huntington's disease. Because Theresa had a 50 per cent chance of having inherited her father's condition, she was denied insurance coverage when her underwriter learned of her father's diagnosis. Even though it turned out later that her father's diagnosis was incorrect,[8] her case highlighted the dangers of genetic testing with respect to insurance cover.

In 1996, the US Congress passed the *Health Insurance Portability and Accountability Act.* To protect against discrimination, it stated that genetic information was not to be treated as a 'pre-existing condition in the absence of the diagnosis of the condition related to such information'.[9] Laws on the use of genetic information for insurance purposes exist in some US States, but not in others. So far, some twenty US States have passed legislation prohibiting the *use* of genetic testing for assessing access to health insurance. The above Act of Congress, however, does not provide any privacy protection because it does not prohibit insurers from gaining *access* to genetic information. Thus, Jackie and Emma—and indeed most of us—have every good reason to be concerned.

In Australia, every permanent resident or citizen has a right to what is largely government-funded health insurance. The problem in Australia, therefore, is not access to health insurance, but is a lack of a formal code of practice for 'collecting, storing, using, disclosing, and disposing of genetic testing information',[10] as well as a lack of formal regulation of the sharing of information among the various health insurance providers.

As for life and superannuation insurance cover, the picture is more complex. Like any other business, insurance companies set out to maximise their profits and minimise their risks. They work on the principle that people pay their premiums according to the risk they bring to the insurance fund. Up to now, the primary risk classification for life insurance coverage has been age. Yet in each age group there are individuals with a lower or higher risk of becoming ill. The insurance companies would, of course, prefer to cover more low risk than high risk people, even allowing for the fact that high risk people often pay higher premiums.

Insurance companies maintain that they need access to genetic information because it would help them to calculate a price that fairly reflects the level of mortality risk posed by a group of people.[11] Genetic tests, they reassure us, would be used in the underwriting process in a similar way to other clinical tests such as cholesterol testing. Already, insurance companies in the UK reserve the right to have access to the results of previous tests. In October 2000, Great Britain became the first country to allow insurers to use genetic tests to identify people with Huntington's chorea. In Australia, insurance companies have put forward a similar proposal. In Germany, too, previous genetic test results may be requested where the policy exceeds a certain amount of money.[12] In general, insurance companies point out that for them diagnostic genetic tests are just another piece of information useful for the underwriting process, that adds to other factors such as age and lifestyle.

In contrast, critics argue that there is a difference between *knowing* one's predisposition to a disease and *disclosing* information about one's medical history. In general, genetic tests are far more accurate predictors of lifespan than a medical history. In addition, since no person is 'responsible' for his or her genome, people should not be penalised with higher premiums for conditions over which they have no control. There is also doubt that underwriters are versed well enough in genetics to interpret and handle genetic information responsibly—particularly where such information pertains to conditions such as cardiovascular disease or breast cancer in which several different genes can be implicated in the development of the disorder. Even if a person has a sophisticated knowledge of genetics no one, at present, can predict with any certainty who will develop what kind of polygenic disease and when.

Is it not surprising, then, that there is a general public drive to prohibit insurance companies to access genetic information? Policy makers in many countries are paying close attention to the arguments on both sides. A clear majority of people does not want insurance companies to have access to genetic data. A public opinion poll in the US conducted for the American Council of Life Insurance (ACLI) showed that 77 per cent of those surveyed in 1994 said that genetic data should not be disclosed.[13] But countries have been slow to legislate, and others have been even more cautious. France and the Netherlands have imposed a temporary moratorium on human genetic testing, while Belgium, Austria, and Norway have opted for an indefinite ban. Even in countries that have moved slowly up to now, the tide is turning against the insurance companies, as public knowledge about the dangers increases.

In Australia, where privacy protection exists at the Commonwealth level, but not necessarily at the State or private level, the Privacy Commissioner took the view:

> that protection [of privacy] could best be extended by giving legislative force to general principles flexible enough to be adapted to the full range of private sector contexts rather than by detailed requirements specific to a particular industry or type of information. Application of the principles to different industries or sectors could be achieved through codes of practice developed in consultation with stakeholders ...[14]

At present, the Life Investment and Superannuation Association of Australia (LISA) follows its own policy on how to handle the use of genetic information and genetic test results. The draft policy states that new applicants are not obliged to have genetic tests for insurance purposes, and aims to ensure the applicant that insurers will maintain confidentiality on all genetic data they may receive. In addition, that insurers will only use the data in the assessment process for that particular individual and not for relatives of the applicant should they also apply. However, the insurer would retain the right to exchange data with other insurance companies that may also be involved in assessing the applicant. In return, insurance companies would ask applicants to provide comprehensive medical information so that they can accurately and fairly calculate the risk of insuring the applicant. In fact, the applicant has a legal duty to disclose that information to the insurer.

While these recommendations are a commendable first step, a number of issues give reason for concern. First, there is the imbalance of a non-binding code of conduct for the insurance industry against a legally binding duty of disclosure on part of the applicant. Second, life insurance companies have no written code of conduct that deals with the appropriate way of how to manage the collection, storage, and handling of genetic information. In the absence of any binding code of conduct, data protection and misuse of genetic information cannot be ruled out.

A last area of concern with regard to genetic testing pertains to workplace practices.

Genetic screening/monitoring & the workplace

For employers, a healthy workplace equates with greater productivity, less absenteeism, and lower insurance premiums. Thus, any genetic information that enters the workplace has both substantial privacy implications and a potential discriminatory influence on employment opportunities and promotion. Employers who become aware of a genetic disability in a prospective employee may, for entirely rational reasons, be reluctant to hire the applicant for fear of that worker becoming ill prematurely or seeking disability compensation. But some critics argue that this is not the only reason that employers welcome genetic testing. They may increasingly embrace genetic testing because of its potential to shield them from claims for negligence and liability in occupational health and safety cases.[15] For example, by selecting and hiring workers who are 'resistant' to harmful substances in any particular work environment, employers may avoid taking remedial actions to minimise environmental risks within the workplace (see also chapter 5).

In the USA and Great Britain, genetic testing has already started to filter down into management and recruitment practices. In 1995, to avoid workplace stigmatisation based on genetic information, the US Equal Employment Opportunity Commission issued guidelines to protect individuals subjected to discrimination where genetic information relating to illness, disease or other disorders had been used against them. Later, members of the Committee on Genetic Information and the Workplace of the National Action Plan on Breast Cancer and the US National Institutes of Health–Department of Energy Working Group on Ethical, Legal and Social Implications of Human Genome Research published a number of recommendations addressing the emerging issues on genetic privacy and discrimination.[16] In brief, they recommended prohibiting the use of genetic information for the hiring or termination of employment; prohibiting requests for or disclosure of genetic information for employment purposes, including conditional employment offers; restricting access to genetic information for employment organisations; prohibiting release of genetic information without prior written consent of the individual—each application for disclosure must inform the individual to whom the disclosure will be made; and strong enforcement mechanisms, including a private right of action, for any violations of these provisions.

In Australia and New Zealand, the indications are that genetic testing has not yet found its way into the workplace to any appreciable level. However, as genetic testing becomes better known and cheaper to perform, this may change. As the range of tests becomes wider and individual tests become more reliable, there is no reason to believe that genetic tests will not become an option for Australian and New Zealand employers.

In Australia, in its *National Model Regulations for the Control of Workplace Hazardous Substances* (of 1994), the National Occupational Health and Safety Commission recommended standards for carrying out health surveillances. These regulations refer to genetic monitoring in the workplace. They aim to ensure that the medical records generated should remain confidential and in the hands of the supervising registered medical practitioner. However, since there are currently no uniform regulatory mechanisms applying to personal genetic information in the workplace, the Australian employee is left with little privacy protection and without the freedom to reject genetic testing should the employer request it.

Is it reasonable to ask perfectly healthy people to undergo genetic testing to secure employment, even though they do not exhibit any physical sign of any disorder, and perhaps never will? Would the applicant not feel pressured into consenting to the testing in order to secure employment, especially in a climate of high unemployment? What happens if the person's job application is unsuccessful? To whom will those data belong? How confidential will those data remain? As one concerned commentator stated:

> Since the employer is often in a position of power relative to the employee or job applicant, unregulated use of genetic testing by employers clearly poses threats to the privacy of employees. As testing becomes cheaper and applicable to a wider range of conditions, the incentives for employers either to obtain the results of previous tests or to seek new tests for their employees will grow.[17]

Conclusion

Genetic testing as a relatively new diagnostic tool is still very limited in its scope and expensive to perform. In this chapter, we could only touch on the most pressing issues concerning its use. We could also only briefly address the wider social context that genetic testing is introducing. But it should be clear that the seemingly unstoppable march of the new genetics is giving rise to profound social, legal, and ethical dilemmas. The new genetics will undoubtedly touch every one of us sooner than later. Jackie and Emma were aware of their vulnerability but chose to act in a manner aimed at saving their lives. One day, it may be common knowledge that each and every one of us harbour genetic mutations that could impact negatively upon our future health. Regardless of that happening, our genetic data ought to remain private, now and in the future. Those data belong to us. They are a significant part of who we are.

Although genes have a significant role in life, they alone do not determine our health, personality, or future. They act in concert with many other factors, ranging from the combined interaction of all genes in an organism to interaction with the internal and external environment. Similarly, genes alone do not determine our social well being. That is shaped by human society and culture. Some groups within society have recognised the immense power that comes from prying into an individual's genetic make-up. Their interest has not remained unnoticed. More and more voices are demanding legislation to protect us from 'gene-hunters' whose actions potentially threaten discrimination in employment and insurance.

The more we arm ourselves with knowledge about the issues of genetic testing and privacy, the better we can respond to any future developments we consider not to be in the interest of humanity. For medicine and for us as individuals, the Human Genome Project promises tremendous benefits. Let us not throw away those promises by allowing the abuse of genetic information about our basic selves.

Knowing your genes: who will have the last laugh?

ROSALEEN LOVE

The public pronouncements that are made about the gene—the gene for this illness, for that aspect of human behaviour—seem to make sense to us. They are given scientific credibility; they are reported favourably in the media (see also chapter 4). They come out of a popular theory about our genetic heritage. Ever since genetics first emerged as a science, it has been widely assumed that genes must influence not only the structure and function, but also the behaviour, of living things. If it is genetic organisation that permits the life of the cell, then genes must significantly determine many distinctively human attributes. Or so it has been imagined. Hence, headlines in the news: the discovery of 'the' gene for alcoholism, obesity, or Alzheimer's disease;[1] there are even notions that there are genes which might promote adultery, be responsible for criminal behaviour,[2] or be linked to high intellectual ability;[3] and there is the overwhelming endorsement of the inevitability of the new science—'Genetics—the future is here now'.

These headlines indicate the gene has not stayed snugly in its supposedly 'objective' scientific place. Now a cultural icon, it has helped geneticists and proponents of biotechnology to produce increasingly powerful images influencing the way people think and speak about themselves. The gene may also be understood as a code or system of meanings.[4] This chapter explores representations of the gene in the contemporary public culture of Australia, highlighting the so-called 'gene for death' and jokes about genes.

By focusing upon the 'gene for death' I hope to capture the paradox of the gene: that while life arises from genes, those genes also contain the potential for both deformity and death. Yet, while nothing in life is more inevitable than death, humans manage to exhibit a remarkable resilience to the notion, with humour a popular way of coping. Equally, in the face of a dominant scientific message about genetic determinism—that we are somehow victims of our biology—there has been a noticeable social resistance through jokes about genes. Jokes about genes indicate that there is a whole other informal debate happening in the public culture in reaction to the notion that we are nothing but our genes.

Genetic literacy

Scientific literacy is a term coined to express the idea that people living in a scientific culture should have some basic understanding of the scientific principles underpinning that culture. It parallels the view that everyone living in a literate culture should be educated to read and write. Two reasons are usually advanced for why scientific literacy is desirable. First, it will help people understand and hence accept the necessity for technological change.[5] The second is the more radical proposal that scientific literacy empowers people to assume an active role in shaping technological change for socially useful ends.[6]

A new phrase—'genetic literacy'—has emerged as a way of conceptualising the duties of the modern citizen in a time of rapid biotechnological change involving significant advances in the knowledge of genetics. Today I need to know my genes in order to be able to deal with the range of decisions I must make, or assessments which may be made of me.[7] These decisions may range from the personal level of genetic screening for family planning and *in vitro* fertilisation, to chromosomal monitoring and genetic screening in the workplace, to DNA fingerprinting in the criminal justice system, to possible genetic insurance clauses. I shall need to make 'genetically informed' decisions about whether I want to eat certain foods, or undergo gene therapies.

If we think of the gene as something fixed—as a fixed unit delivered through time to new generations—we can trace a path of inheritance. I look at my hands, and I see my mother's hands. I imagine this as a gift from her, given to me as genes. I imagine she, in turn, received this gift from her parents. I look at my feet, and I see my father's feet. I imagine this gift from him, from his parents to him, and so on back through the generations. I imagine the segment of DNA— the gene—as something fixed, which is handed along as a gift. From where did this gift arrive? How far back must we travel? Initially I imagine human characteristics, but I soon learn this is rather biased to a human-centredness viewpoint. Scientists say we share something like 99 per cent of our DNA with the primates, 90 per cent in common with the rat. The bacteria, the virus, and the common rat have something in common with my genome.

In the history of genetic diseases, the gene is, again, presented as a fixed unit of inheritance. Genetic diseases are particularly awful in the transmission from unknowing parent(s) to a child. Huntington's disease, Tay-Sachs disease and other diseases which degenerate and kill the recipient, horribly, either late or early in life are examples here. In reflecting on the history of disease, I contrast genetic disease, the often fatal internal flaw, with epidemic disease, the bacterium, the virus, which strikes from without. Yet the two are linked. Humans have survived attacks by viruses and bacteria. But they have survived not just by conquering the infection, 'throwing it off', as we say, but by incorporating part of the invading organism *within* the human genome. One of the fascinating insights from recent work in immunology and genetics is that gene fragments from organisms that caused previous devastating plagues like the Black Death have been found in the genetic make-up of modern humans.[8]

Miroslav Holub (1923-1998) was both an immunologist and poet. The poet in Holub marvelled at the genome—'the logical record of an integrated organism's inner evolutionary drama'.[9] I may start with imagining my genes as a gift from my parents, but I find I possess a genetic gift from all life on earth. Holub saw the inner evolutionary drama played out in the genome. The genome is, he argued, using the notion of DNA as inscription in the Book of Life, 'a genetic chronicle, a good fifth of which is written in absolutely primitive, viral syntax'.

I now imagine the human genome as something of a conquistador—opportunistic, entrepreneurial, a take-over merchant in the battle against microorganisms that treat humans as their prey. Some swift deals have been made in the genetic past. My respect increases. 'It is no co-incidence', Miroslav Holub continued, 'that some of our (relatively noble and relatively human) factors and intercellular signals have nucleotide sequences very similar to retroviruses'. Former predators lie tamed within.

I began by imagining the gene as a fixed unit of inheritance but have found there is a certain amount of fluidity in the notion of 'fixity'. To this I must add, as I explore the notion of genetic literacy, the concept of 'mutation'. The gene may be subject to change, either as the result of an error in the process of replication, or as a result of environmental factors. Mutation is defined as any replicable change in DNA sequence. A good gene changes to bad, replicates, and magnifies the error. Genes for particular kinds of cancer are being talked about in two ways; the gene may be a 'bad gene' in that it is inherited as a gene for a particular form of cancer (breast cancer and colon cancer figure largely in these narratives); or it can be a 'good gene gone bad', a gene inherited as harmless, but which suffers a mutation and becomes cancer producing.[10] The good gene goes bad and takes control of its immediate environment. Cells multiply, soon out of control. Malignant tumours grow because cells, which normally have a death switch, cannot turn that death switch on—they grow out of control, and the sufferer often dies.[11]

The gene for death

Each day it seems as if a new gene for this or that disease is being proclaimed in terms which reinforce the sense of genetic destiny. People speak out about difficult personal decisions made in response to the news that they bear a 'bad' gene. Social workers become genetic counsellors. Different groups have entered the debate about genetic issues. For example, the Central Australian Aboriginal Congress refused to participate in the human gene collection activities of the international Human Genome Diversity Project—redefined as the 'Vampire Programme' by indigenous peoples; the Federal Privacy Commissioner has inquired into protection for genetic information collected in medical tests and criminal cases; the Commonwealth Department of Environment hosted roundtable talks in 1994 on 'Access to Australia's Genetic Resources', still unresolved at the time of writing. Yet what 'gene' is it that is so often discussed in these various contexts? The gene is rarely given a scientific definition in these public debates. Rather, it is a term to which participants bring diverse and everyday interpretations of meaning. In the near future, however, the phrase 'knowing your genes' will take on a new urgency with the production and consumption of 'gene-test products'.

'Gene tests. How you will die' were the large words on the street publicity-flyer of *The Australian* on 8 August 1994. The article's caption 'Genetic Tests put Fate up for Sale' (p. 3) revealed that knowledge of how you will die will soon be available for purchase in the marketplace as gene-test commercial products. Between 1994 and 1998,

DNA-based gene tests became available for 30 diseases. By August 2000, the genetic testing resource *Genetests* listed 769 diseases in its directory.[12] 'Knowing your genes' means knowing how you—or your children—will die. For example, with the cystic fibrosis test, potential parents can determine what chance their future children have of inheriting a double dose of a recessive gene for what is termed 'the most commonly inherited killer disease'. Socially, the genetic destiny transition is from 'how you will die' to 'how through abortion you may prevent a killer disease in others', highlighting that gene-testing technology does not lead to any 'easy' solutions (as Rowland highlights in this volume). Issues of choice, ethics, social responsibility and moral beliefs are opened up—all of which are beyond the 'fixity' of genes.

The paradox between public pronouncements of the notion of genetic destiny and the social reality of treatment has often led to a noticeable use of disclaimers from scientists. Take the comment from Dean Hamer (now Chief of Gene Structure and Regulation research at the US National Cancer Institute) in publicity for his discovery of the 'gay gene' (reported in the 1994 ABC-TV *Four Corners* programme); 'many people have the idea that genes are destiny and that genes are some sort of master-puppeteers that are jerking our strings, sexual and otherwise ... that's a very incorrect view of how genes act'. This disclaimer came after a number of images had seemingly pointed in the opposite direction, that is, towards genetic destiny. Two gay brothers were shown giving blood for Hamer's project. They looked like brothers. The marker pen pointed to bands on strips of x-ray film—the bands from the gay brothers appeared to coincide. The narrator said 'we think that part of the [X] chromosome contains a gene that influences their sexual behaviour'. The marker pen pointed to the spot. The viewer saw it. Both brothers have it. The message is, genes are destiny, even if the scientist does not himself here make the strong claim that having the 'gay' gene causes homosexuality.

Again, what is understood from media reports of Hamer's research is often more than is said. Reports are interpreted as the discovery of the 'gay gene', the 'gene for homosexuality'; what is actually said on the Hamer programme is something less than this. Reference is made to 'this portion of the X-chromosome which the two brothers have in common', and 'we think that part of the X-chromosome contains a gene that influences their behaviour', but the message seemingly delivered is that 'the gene for homosexuality' has been discovered.[13] 'Knowing your genes', for a gay person, now means what? Comfort for some; for others, indignant rejection of assertion of genetic determinants for behaviour freely chosen. American biologist Garland Allen suggests that instead of talking about *the gene for* some factor, the notion of *the gene that has a norm of reaction for* a particular factor should be substituted. This takes into account the complex relationship between the gene and the environment in which it may, or may not, be expressed—a complex relationship to which Dean Hamer was alluding in his comment above.

Saying the gene is destiny helps define the issue in a certain way, a way that may be perceived quite positively. 'Designer destinies' is the title of an article in the *Economist* (1994), subtitled: 'the first steps towards human genetic engineering give no cause for alarm'. To see genes in terms of 'designer destinies' acknowledges and promotes the notion that there may be potential economic benefits from gene therapy. The issue of 'death and the gene' is redefined in a way that promises some hope rather than no hope. 'Knowing your genes', in this case, brings new choices.

Dissociating the issue of genetic destiny from negative themes is only possible up to a point. The biggest negative theme of all for humans is, clearly, death. What is new about recent pronouncements about the gene and cancer is the degree of certainty of the statements coming out of cytobiology and cytogenetics. Information about 'the gene for' this or that kind of cancer is being given in much more sharply focused, much more dogmatic form. If you have this gene, the headlines report, you will be 99 per cent certain to contract this form of cancer. It will be 'virtually inescapable'.[14]

Particular families are profiled. *Time* (in its Australian as well as its US edition) printed a photograph of American Anna Fisher, which showed her surrounded by photographs of her mother and other female relatives who have all died from breast cancer. 'Malignancy is simply part of her pedigree', the article comments. Fisher recovered from ovarian cancer. Then she was given, and took, the option of a prophylactic double mastectomy;[15] that is, she did not have any signs of breast cancer at this time, but she chose to avoid any chance of getting it in the future by having her breasts removed. With the pictures of those who had died, as well as the stark—supposedly 'rational'—choice made by the survivor, this material is both evocative and extremely powerful.

'The Big Killers', a table in the *Time* article, listed estimates from the American Cancer Society about thirteen forms of cancer and their mortality and survival rates, and risk factors. Such a list is, however, much more than summary information simply transmitted to the reader.

Against the items on this list the reader may place the name of a person: my mother, my cousin, my friend and, potentially, myself. In this scenario science becomes, as Carey observes, 'a world of dramatic action in which the reader joins, a world of contending forces, as an observer at a play'.[16] But the underlying tension is that we are not only observers but, being mortal, are also participants.

Does knowledge of which genes might kill us provide us with new freedom, new options? Genetic therapies are not yet available, and this is one of the problems with the release of information about 'bad' genes. But now women like Anna Fisher have the freedom to make a grim choice of prophylactic double mastectomy. It is a freedom to choose, rather than freedom that is granted when restrictions (for example, genetic restrictions) are removed. The restriction, the bad gene, is still there. Anna Fisher knew it in a way her mother did not. She had a consumer choice. She had knowledge to help her make a hard decision—and made it. The term, 'the gene', imposes order on biological existence. It grants a new form of identity, a genetic identity to the individual. It helps inform narratives of self, providing a sense of something solid and basic, however unwelcome, to the understanding of 'how you will die'. 'I am bound, yet I am free'. The conditions of my being bound, the new genetics and the scope of my genetic literacy, are also the conditions of my new freedom.

Genetic literacy & the flying pig

The term 'genetic literacy' embraces a number of contrasting perspectives. One perspective is unashamedly science-centred, and relates to what scientists do to convince the public to believe in, and co-operate with, the process of (bio)technological change.[17] This perspective works as a one-way transmission model of communication. Here, science produces a certain field of knowledge and then informs the public, which the public then takes and acts upon to make 'informed' choices or so-called 'rational' decisions.

An example of science-centred communication was a Commonwealth Scientific, Industrial and Research Organisation (CSIRO) travelling exhibition on genetic engineering that toured the Westfield shopping malls of Australia in 1992–93. The exhibit was crowned with a series of flying pink pigs—porco-avian icons subtly aimed to ridicule critics of biotechnology. 'Of course pigs will not fly as a result of the new biotechnology' was an explicit message from the exhibition, intended to reassure those with doubts about genetic engineering, and to convince others about its attractions (Hindmarsh also remarks upon this in chapter 2). The CSIRO team sought to instil genetic literacy and provide reassurance with a mix of facts, words, personalities, concepts, history, and ethical implications. Interactive skip videos allowed a level of public participation. At the end of each video sequence, viewers were asked to record their vote on a controversial issue, such as 'who should control genetic engineering?'. The tally of votes was then displayed, allowing participants to see where their vote fitted in the public response.[18]

A second perspective on genetic literacy arises in resistance to the science-centred stance. From this perspective, scientific communication, if it is uncritical in its praise, may send an unintended message to the receiver that science is above human scepticism.[19] A typical response may be 'well they would say that, wouldn't they?'. Or, as I suggest later, if a dominating sense of genetic essentialism—the notion that the gene marks the essence of human identity—is conveyed, resistance might occur in the form of jokes about genes.

We can also ask other questions: what might genetic literacy look like, twenty years into the future? What will genetics mean by then? Where will people turn for genetic information and advice? What will motivate them to do so? Some of the answers might well alarm the scientists. What if a New Age public gives as much scientific legitimacy to the so-called 'intuitive sciences'—palm-reading, astrology, clairvoyance—as to genetic engineering? What if, instead of official biomedical science, alternative or genetic therapies are sought? Genetic literacy might then become something very different from TV viewers trying to interpret the meaning of signs on DNA marker gel to workshops on getting in tune with your DNA.

How popular culture has constructed and pushed a certain image of the gene was recently tackled by Dorothy Nelkin and M. Susan Lindee in *The DNA Mystique: The Gene as Cultural Icon*. Nelkin and Lindee collected examples of conversation about genes from sources in US supermarket tabloids, soap operas, parenting manuals, biographies of Elvis, and more. They argue that the images and narratives of the gene in US popular culture reflect and convey a message of genetic essentialism. They see these notions as dangerous, lending themselves all too easily to misuse in the service of socially destructive ends,[20] especially through genetic discrimination. Within the world of genetic essentialism, 'knowing your genes' is tantamount to knowing your biological constraints and thereby accepting a certain biologically allocated place in society.

Jokes about genes

Notions of genetic essentialism as a dominant cultural notion has met (expectably) with resistance. One form of resistance is found in satire. Jokes about genes position themselves against the prevailing genetic essentialism. People telling jokes are taking a sceptical look at the messages of science mediated through popular culture. They are saying, 'Oh yes, and what else?' or, more cynically, 'What a load of rubbish'. In response to being told one's place, told to accept what others deem good for us, one reaction is to poke fun at the new technology and those promoting it. Jokes about genes indicate a certain scepticism especially directed to the institution of medical science, and news broadcasts which report new miracles of modern genetics in the context of global problems, crises in health care, poverty, famine, and war (see also chapter 1).

In one episode of *Kittson and Fahey* screened on ABC-TV on 1 November 1993, Mary Anne Fahey and Jean Kittson are shown puffing away on cigarettes. The scene is a factory shed, framed in a smoky haze. Mary Anne confides to Jean that she cannot help it. She has the gene for smoking. In 1993 I laughed at the absurdity of the 'gene for smoking'. But by early 1999, scientists found three genes they claim shows an effect on smoking, with one version of the gene thought to discourage nicotine addition. Behavioural genetic studies, Dean Hamer claims, show that cigarette smoking is 53 per cent heritable and that there are different genes for starting and continuing to smoke.

However, another episode of *Kittson and Fahey* provides a more enduring example. The mad scientist is here the mad genetic engineer. Jean tells Mary Anne about the doctor Mary Anne is about to consult: '[He] is into gene shearing, impregnating women with the genes of goats, sheep, pigs. Breeding a woman that could cook, clean, and bring home the bacon too'. Mary Anne is deeply worried. In the 1993 stage show, *Look at me when I'm talking to you*, Dame Edna Everage claims she is going to have a baby, but not in the usual way. It is growing in a saucer on the doctor's window ledge. It all dates back to when Dame Edna was staying with her doctor, and found in the fridge a phial with a use-by date. The name on the phial was Arnold Schwarzenegger, and the contents a product of Arnie's early days as a struggling young sperm donor 'before the germinator became the terminator'. Dame Edna has fond hopes for her offspring: 'It will have my body, my brains, and his income'.

The beauty of the throwaway line is that it may cleverly compress pages of logical argument into one telling, funny phrase that will remain in the mind for a long time. Masters of this trade are Australian comedians H.G. Nelson and (Rampaging) Roy Slaven. Roy is cheerfully omniscient, always ready to comment on any topic from life on Mars ('Call *that* life on Mars?' to the medical status of a footballer's groin. To celebrate the twentieth anniversary of the death of Elvis, Roy and H.G. produced on one ABC-TV *Club Buggery* episode some Elvis memorabilia for their audience. Roy brandished a toothpick Elvis had allegedly used. From this toothpick, Roy claimed he could extract Elvis's mitochondrial DNA. H.G. Nelson was suitably impressed. 'Clone Elvis?' he asked. 'Clone Elvis? Clone his clothes? Clone his guitars?' Roy nodded. H.G. continued: 'Which Elvis would you clone? Young Elvis?' 'Of course not', said Roy. 'I'd clone big fat Elvis'. The viewer is left with an awesome vision of what the world would be like, awash with cloned fat Elvises.[21]

Cartoons and pictures provide another base for clever 'deconstruction' of the experiments of scientists. Scattered throughout the New Zealand text *Designer Genes* are numerous cartoons of the world of the future. In one the insertion of the toad gene into a potato is represented by a rather ugly 'potatoad' with legs, but a sprout for a head. In another, a customer is sitting at a restaurant table in front of a foaming plate of food complaining to the waiter 'I ordered quiche!'. The waiter's response is, 'This food is substantially equivalent to quiche. I'm sure you will like it'. A drawing of an udder producing tablets instead of milk is titled 'pharming'.[22] A slightly different theme was explored in *The Ecologist*. After examining his patient the doctor explains, 'I'm going to put you on a course of hormones—I recommend drinking three pints of milk a day'. A second cartoon in *The Ecologist's* feature about bovine growth hormones has a mother and father at the farm gate berating their son for having gone unaccompanied into a paddock. The child's right foot has grown ten times larger than the left, with the caption saying 'Trust you to go and stand in some growth hormone'.[23] Finally, a rather disturbing pig-faced tomato stared from the plate at readers of the Focus section of the *Weekend Australian* in May 1999. The caption read 'Jellyfish in sugarcane. Flounder genes in strawberries. Human genes in pigs ... you are what you eat now has a new meaning'.[24]

Jokes about genes and biotechnology provide a bridge between popular and scientific cultures. Games are played with scientific ideas, with both scientists and non-scientists mocking scientific hype. Jokes about genes rise above the charge often levelled at satire, that it uses mockery in defence of a conservative *status quo*. Claims made about the genetic basis of what it is to be human are recognised as very important. But if the message is that the causes of major diseases and behavioural problems are 'in the genes', and if the claim is then made that fixing the genes will fix the problem, sceptics might be expected to respond with jokes. The jokes point to the gap between grandiose claims that science has the answers and the everyday world of technological risk and hazard where scientific certitude is being increasingly questioned.[25]

Again, laughter enters through the gaps in the wall of facts that science has proudly built. What Nelkin and Lindee demonstrate so convincingly in their book is that the gene exists as much as a cultural creation, a cultural icon, as scientific entity. In his narrative of adoption Robert Dessaix said of himself, when he found his biological mother: 'I realised that all sorts of things I had carefully crafted and constructed in myself were actually given to me.'[26] This gift is increasingly being labelled 'the gene'. The gene takes on a range of cultural meanings that the hardworking laboratory geneticist must blanch to consider. Dessaix reports meeting a room full of his new-found relatives, and finding both he and they shared a common quirk: 'I thought I jiggled my knees because I was stressed or nervous. It turned out everyone jiggles their knees. From here to the horizon they are all knee jigglers.' The humorous observation, and the joke, take off precisely because there is no one meaning we all agree to give the term, 'the gene'. There is paradox and inconsistency in the information that we are given. Learning to live with ambiguity is part of the process of getting to know your genes.

Conclusion: the human genome & the human spirit

I referred earlier to the words of Miroslav Holub, the Czech immunologist and poet. When asked how he saw the relation of science to religion, Holub replied:

> I couldn't say whether I am religious. I would obviously be as a unit,
> as a poor unit which is just an epiphenomenon of something bigger.
> The something bigger I would rather describe as a genome and not

as a spirit. But anyway, it's something supra-individual; the genetic process of the planet. We are obviously in the position of religious individuals because it is way above our heads and we are not the aim of the process.[27]

Holub went beyond the here-and-now of his individual life to see it as part of an evolutionary whole which stretched from the past history of life on earth into its future. He talked about the relationship of part to the whole of life, not, he stresses, in spiritual terms, but 'as a genome rather than a spirit', or 'a sort of instinct for survival—a deep homeostasis in human life'. The nineteenth-century notion of the Life Force is here demystified to the Life Thing (or as the selfish gene)—the entity DNA. Yet, as biologists displace themselves further from nature, through computer-aided analysis of gene relationships, Holub yearned for something more, to somehow experience the relation of the part to the whole as best he could in his life work. He placed himself before the genome and the genetic process of the planet in a spirit of humility: 'we are not the aim of the process'. This something bigger, 'the genome', is interpreted as a holistic entity with which scientists may co-operate in extending human understanding. Holub experienced the spiritual feeling of awe in the presence of something, the genome, that transcends the existence of the individual caught in a particular time and place.

Holub read the parts of the DNA we share in common with other mammals as 'the logical record of an integrated organism's inner evolutionary drama'. In the battle between life and death, a compromise has been reached in which life continues by appropriating its former enemies to its own ends. In interpreting the history of life on earth in terms of the conquistadorial activities of genomes, Holub found acceptance of the fact that the death of one individual is often necessary for the life of others: 'Tragic death is a precondition of biological optimism.'[28] He imagined a biological or genetic supraconsciousness to which the question of individual death is not central. The biologist must find meaning in accepting responsibility for the planet as a whole and as a viable system. Viability thus takes on planetary dimensions and includes the preservation—not only of human civilisation—but of life in general. Holub questioned whether humans have the wisdom to recognise this.

Holub's writings concur with warnings from ecologists and others who have studied environmental degradation. For example, Wes Jackson believes that the sustainable agriculturalist must begin with the idea that agriculture cannot be understood on its own terms. Agriculture arises from nature. He poses the question: Should a crop plant be regarded as the property of the human or as a relative of wild things?

> If it is viewed primarily as the property of the human, then it is almost wide open for the kind of manipulation molecular geneticists are good at. If, on the other hand, it is viewed as a property of nature primarily, as a relative of wild things, then we acknowledge that most of this evolution occurred in an evolutionary context, in a nature that was of a design not of our making.[29]

Both Jackson and Holub urge us to remember this last point—that evolutionary patterns in nature are of a design not of our making (as Wills emphasises in chapter 3). Yet, scientists are daily assuming otherwise, acting as if genetically engineered crop plants, or the cloned sheep 'Dolly', were wholly the property of humans. Holub, in suggesting a notion of some kind of supra-individual process of the planet, is going beyond the here-and-now of biological knowledge, but doing so in a spirit of humility. While remaining first and foremost an immunologist who valued his work to alleviate human suffering, Holub took us beyond the materialistic assumption that the gene is merely another object to be manipulated. For Holub, knowing our genes meant also acknowledging our ignorance about our genes.

It must be very exciting for scientists working in the new genetics. So much is happening, and so quickly. As David Suzuki has discussed so vividly in the introduction to this book, research students just out of their postgraduate degrees find themselves in research teams doing manipulations undreamed of when their professors were young. Information is now 'tumbling off the walls', as one excited scientist remarked. Yet, because there is so little place for ethics and the social and broader ecological context in science education, the tendency is to apprentice the young scientist to explore science with a model of genetic determinism which seems to work well at the practical experimental level. What these genetic manipulations mean to the everyday lives of people is thus a topic remote from the laboratory, and when it does come up it usually does so in the context of science-as-progress—we will all ultimately benefit from genetic engineering.

But it is the unintended consequences of these well-intended human actions that are the great unknowns in this story. The comic throwaway line, 'Clone Elvis?' raises a spectre of delight for some, pure horror for others. One thing is certain. Twenty years from now there will be genetic applications few outside the lab have yet to imagine. They will potentially affect everyone's lives. People may be joking now but tomorrow the joke might be on us. That is why this 'knowing your genes' and the social context of genetic literacy is so important. Not only can such knowledge improve one's ability to question and to resist manipulation, but it can also enhance public debate and participatory decision making in this most uncertain and profound area of technological change.

8

Risks, regulations & rhetorics

STEPHEN CROOK

This chapter analyses the 'riskiness' of biotechnology, the ways in which its proponents and opponents articulate that riskiness, and the ways in which riskiness is managed. We usually think of risk management as something to do with calculations performed by scientific experts and with regulations put in place by governments. However, there is another less obvious site at which the risks of biotechnology are managed: the site of rhetorical struggles over the cultural meaning and what we might term the cultural riskiness of biotechnology. For its opponents, biotechnology is a source of novel and monstrous hazards linked to a fundamental shift in the human relationship to nature. For the proponents, by contrast, biotechnological innovations are modest, natural, and productive improvements, the latest in a long line of benevolent human interventions in horticulture, animal husbandry, medicine, and other fields.

The stakes in these struggles are very high. Unless the proponents of biotechnology can prevail, they will not convince the public that any amount of expert calculation or government regulation of risk will secure the safety of biotechnological processes and products. On the other hand, the best chance for opponents who want to build public resistance to biotechnology is to convince the public that biotechnology is culturally risky as a whole. Just to complicate matters, the rhetorical struggles over biotechnology risks cannot be wholly disentangled from the risk assessments and regulations of experts and governments. It may sometimes seem, for example, as if regulations are put in place as part of a rhetorical strategy to reassure the public rather than for any more precise technical effect.

These issues are explored below in five sections. The first uses the idea of a 'risk society' to explain why we have become so preoccupied with risk in recent decades. The second explains why biotechnology is 'culturally risky'. The third discusses the links between the scientific control of nature and the ways we try to manage risk. The fourth reviews the rhetorical strategies through which proponents of biotechnology manage its cultural riskiness. The concluding section considers the current balance of forces in the rhetorical struggles over the cultural meaning and riskiness of biotechnology.

Life in a 'risk society'

The German sociologist Ulrich Beck claims that in the advanced West we have passed beyond an 'industrial society' to a 'risk society'.[1] Society is no longer built around the production and distribution of goods but the production and distribution of 'bads'—chiefly, environmental and technological hazards arising from advanced industrial capitalism. These new risks are present everywhere, but we rarely experience them directly: we cannot taste the pesticides on our fruit and vegetables, smell the pollutants in acid rain, or see the build-up of carbon dioxide in the atmosphere. However, their consequences are increasingly visible in unhealthy children, the death of the German forests, or extreme weather events.

British social theorist Anthony Giddens has similar ideas about risk. He argues that our present stage of social development imposes upon individuals unparalleled demands to make choices and to monitor their own behaviour, but in circumstances where they are also radically 'disembedded', or uprooted, from the institutions—family, community, class—that once furnished trustworthy recipes for the conduct of life. In the place of these older social institutions, complex 'abstract systems' have grown up that order our lives in the advanced societies. Examples include money and the financial system, arrangements for medical care, the generation and distribution of electrical power, computer networks and the networks through which food is produced, packaged and distributed (see also chapter 9).

The generally smooth operation of these systems reduces the likelihood of adverse outcomes, but raises the stakes on system-failures when they do occur. So contemporary mass production and distribution systems for food may be more hygienic than the localised practices of the past, but the consequences of a single failure—contamination of a batch of processed meat or peanut butter, say—can be great. These low probability but high consequence risks are part of a 'sombre side' of modernisation (along with ecological damage, totalitarianism, and war). They promote a 'risk climate' in which risk anxieties become constant preoccupations, according to Giddens.[2]

To summarise, we have become preoccupied with risk for three main reasons: because we are increasingly confronted by the hazardous by-products of advanced industrialism, because we face these and other hazards with few communal supports, and because we are becoming ever more dependent on abstract systems that generate low-probability but high-consequence risks.

There are direct connections between these arguments and current anxieties about biotechnology. For example, fears about the invisibility of extensive and potentially high-consequence risks underpin debates about the labelling of genetically modified (GM) food. Two British supermarket chains (Asda and Iceland) banned unlabelled modified foods in 1997. According to press reports, Iceland was concerned 'that a government decision to allow the import of soya, some of which is genetically modified, means that it could be used in processed foodstuffs without being traced'.[3] A molecular biologist was quoted in the same story as saying that genetically modified foods 'could give rise to the unwitting spread of new poisons, new allergies and even a reduction in a food's nutritional value'.

Similar themes emerged in Australian coverage of the same 'gene bean', a Monsanto product tolerant of the broad-spectrum herbicide Roundup (another Monsanto product). The policy officer of the Australian Consumers' Association protested that the bean had arrived 'willy nilly' with other soya beans, and could find its way into 60 per cent of supermarket foods.[4] The general concern that lay behind these specific complaints was formulated by a British 'prize winning micro-biologist' in terms that were very close to those used by Beck: 'Introducing a gene into another organism is a Russian roulette process—the position the new gene occupies is not controllable. We are being asked to partake in a nutritional experiment of global proportions.'[5]

Proponents of biotechnology were irritated and frustrated by such claims. British government scientists complained that 'ill-informed "scaremongering"' about GM foods had produced unwarranted consumer resistance in Britain.[6] Unwarranted or not, consumer resistance continued to grow in Britain, culminating in 1999 in an almost complete collapse of support for GM foods and agricultural biotechnology. In Australia, the official response to the 'gene bean' controversy from the food industry's representative body at the time, the Australian Food Council,[7] was very similar to the UK authorities': the beans constituted 'no public health or safety hazard whatsoever'; labelling would be 'inappropriate, impractical and meaningless'; and the gene bean had 'the same composition, nutrition and processing characteristics as conventional soybeans'.[8] However, in a sign that consumer resistance to GM products has grown in Australia, too, the Council's successor organisation has softened its position. A press release late in 2000 asserted that the Council 'has consistently advocated strong GM food safety assessments and labelling laws, that protect public health and safety and provide meaningful information for consumers'.[9]

The cultural dynamics of risk

While all human societies in all historical periods have faced dangers and hazards, the idea of 'risk' is definitively modern, according to Beck and Giddens. That is, 'risk' is to be understood as the systematic and future-oriented way in which modern societies process dangers and hazards by weighing them against opportunities. This modern and dynamic concept of risk is contrasted with traditional orientations to fortune, fate, and destiny.

Such sharp distinctions between the traditional and the modern have been contested by the American anthropologist Mary Douglas.[10] Douglas insists that all societies develop cultural, social, and political mechanisms for processing perceived hazards, mechanisms that reflect both the wider social structure and the way the society has selected particular risks for special attention. This is true of Western societies that worry about environmental pollution just as it is of African societies that worry about witchcraft.

It is important to note that Douglas' approach understands risk as something that is always relative to a particular society and culture, rather than something that is objectively 'out there'. For example, it has been argued that the citizens of ancient Rome suffered high levels of lead poisoning from the channels and pipes they used to distribute drinking water. In Douglas' terms, lead poisoning was not a 'risk' for Roman society because it was not recognised, let alone managed, in any way. By contrast, witchcraft was identified as a serious threat in many European societies in the early modern period. Whatever our views about the 'reality' of witchcraft, it was a risk for those societies in Douglas' sense because it was identified and managed as such.

The way in which a given society selects risks for attention and attempts to manage them is an important part of the way it gives itself a clear cultural identity. It is here that Douglas' work can directly help us to understand why GM foods and biotechnology more generally are culturally risky. One of her interests has been the way in which many cultures define themselves in relation to risks of pollution.[11] Witchcraft, wrongly prepared food, menstruating women, and particular animal species have all been 'selected' by some culture or another as a fundamental threat to its 'purity'. Pollution threats arise when boundaries that define a culture are breached: for example, when non-food species are eaten, when the milk is mixed with the meat, when humans meet with spirits. This conception of pollution connects with the principle that cultures are built from basic binary oppositions: nature–culture, male–female, raw–cooked.

On this basis, biotechnology can be understood as culturally risky because the very principle of the genetic engineering of plants and animals breaches sets of distinctions between the 'natural' and the 'artificial' that have long been established in Western cultures. Those distinctions, in turn, are built into a complex web of commonsense assumptions about pollution, risk, and safety. Campaigns in the US and Europe against the use of recombinant-DNA (r-DNA) derived milk-boosting growth hormone somatotropin (bST) in milk production are opposing a kind of cultural 'pollution' when they emphasise the natural 'purity' of milk and the threat of 'contamination' posed by bST.[12]

Non-modern societies often use myths and folk-tales to explain the origins and consequences of pollution: the marriage of Coyote to Yellow Corn Girl brings witchcraft to the Native American Pueblo; and in an ancient Greek myth, Pandora releases disease and pain into the world when she opens her eponymous box. Modern societies, too, have their mythic representations of the dangers of mixing the natural and the artificial, from *Frankenstein* to *Terminator* and *Jurassic Park* (as chapter 1 also points up). To the British press, genetically engineered crops are 'Frankenstein food'. The cultural riskiness of biotechnology, understood in this way, is at the heart of rhetorical struggles over biotechnological risks, and will be considered in more detail shortly.

Control, risk & risk management

Modern science has been characterised as a long-term project aimed at human control of nature. Its success has been linked to the 'reductionism' of science—its search for the basic component parts of natural phenomena. Once these parts are identified and the relations between them specified, larger-scale processes can be controlled through the manipulation of the parts. Physics is the model for this project and, on some accounts, biology did not begin to match the pattern until the emergence of molecular biology in the 1940s. Before then, the work of observation, classification, and speculation that occupied biologists made their field look like a 'third world scientific territory' to physicists. Physicists such as Schroedinger and Slizard played a major role in the development of the new molecular biology, which undertook 'the investigation of the ultimate units of the living cell in the same way that physicists and chemists investigate the ultimate units of matter'[13] (see also Wills' discussion in chapter 3). In molecular biology, as in other reductionist sciences, there is a close link between analysis and control. The idea of identifying the 'ultimate units' of the living cell carries within it the idea that those units might be manipulated and rearranged. It is a short step in principle, if a long march in experimental practice, from Crick and Watson's 1953 model of the 'double helix' to genetic engineering.

As a project for the control of nature, science relates to risk management in three main ways. First, science and technology provide tools for the direct, or first-order, management of natural hazards. From scientific medicine to power generation to weather forecasting, science and technology are at the heart of many 'expert systems'. Second, the ways in which risks are identified and assessed have themselves become increasingly 'scientised'. For example, the forms of risk insurance—life, medical, automobile—which we take for granted could not have become established until the mathematics of statistical sampling was systematically applied. Third, as Beck argues, many environmental risks are now 'second-order' risks—risks that arise from the attempt to control other risks. For example, a new medical treatment, a nuclear power plant or a genetically modified organism may have hazardous consequences that are worse than the hazards the innovation is designed to control. To ensure that this is not the case, experts conduct risk assessments and trials before the medical treatment is recommended, the power plant built, or the transgenic organism released into the environment.

An important consequence flows from these connections between science, risk, and control. The whole complex task of identifying, assessing, and managing risks can often appear as a purely technical matter, best left to the experts. Returning to the example of the Monsanto 'gene bean', who was to assess the risk posed by this product of the plant geneticist's craft but other plant geneticists? Hence, the exasperation of the Australian Food Council in the face of complaints about the unlabelled release of the bean: the competent experts had declared the bean safe because it was 'substantially equivalent' to other beans. Objections could only have been ill-informed, irrational or mischievous. Irritation at the ignorance and irrationality of those who cannot grasp the basic principles of expert risk assessment is widespread among risk specialists and promoters of new technologies.

Governments charged with the responsibility for regulating and managing risk find themselves in a quandary on this question, torn between public disquiet and expert advice. An important current of US opinion holds that risk management has been too 'public-opinion based' rather than 'risk based', and that this must change. As Zechauser and Viscusi put it, 'government policy should not mirror citizens' irrationalities but ... should make the decisions people would make if they understood risks correctly'.[14] At the other end of the spectrum stands Beck's cynical view of expert risk assessment: because science bases risk assessment on case-by-case, laboratory-based experiments, and can never hope to match the complex interactions of the 'real world', such scientific assessment is limited and ultimately fraudulent. Science has become 'the protector of global contamination of people and nature ... the sciences have squandered until further notice their historic reputation for rationality'.[15] This may be an overstatement, but over recent years governments and biotechnology corporations have been forced to face up to a dramatic collapse of public trust in politicians and 'experts' of all kinds. The implications of this collapse are considered below.

The whole burden of the discussion of Douglas' work in the previous section was that pure calculations of risk—uncontaminated by culture—are not possible: there is an irreducible social, cultural, and political dimension to risk. To give that point a more specifically sociological twist, the identification, assessment, and management of risk can take place only within patterned policy settings and institutional arrangements—what might be termed 'regimes of risk management'.[16] Over the past thirty years or so, the typical regimes in place in Western societies have been subject to unsettling change. The regulation and eradication of risk has been a major element in the programmes of what Beck has termed the 'provident state' and others, the 'welfare state'. With the support of major interest groups, such states put in train a regime of 'organised' risk management. Legislators, experts, and enforcement agencies co-operate in the systematic identification, assessment, and management of risks—ranging from industrial pollution and infectious disease to unemployment and illiteracy.

Over the past twenty years or so the bureaucratised and centralised structures of the 'provident state' have come under attack, so that Australia is now well advanced down the road of a 'neo-liberal' redefinition of the role of the state. The symptoms are familiar and include sales of public assets, financial and trade de-regulation, labour market reform, the downsizing of the public service. In risk management there is a corresponding scaling-down of the organised regime. In areas from emergency services to air traffic control and meat inspection, there is a trend towards industry self-regulation and the contracting out of services. Perhaps most critically, there is a heavy emphasis on individual responsibility. Individuals are called upon to govern themselves, to secure their own welfare through informed calculations of risk. In the emergent 'neo-liberal regime' of risk management, state services are transformed into bureaux for the provision of information and expert advice to 'responsible' individuals and industries.

In Australia, the European Union, and the United States, biotechnology is subject to the bureaucratic surveillance and regulation associated with the organised regime. However, arrangements for its regulation have often emerged in an *ad hoc* way, shaped by uncertainties and conflicts about where biotechnology fits into existing bureaucratic structures. Australian arrangements are presently in transition, with a new and simplified structure established in three Bills brought before the federal parliament in mid-2000.

Under the new arrangements, an 'Office of the Gene Technology Regulator' (OGTR) located in the Department of Health and Aged Care will take overall responsibility for the provision of advice to government and the oversight of regulations. In turn, the OGTR will be advised by three bodies. The Gene Technology Technical Advisory Committee (GTTAC) will replace the existing Genetic Manipulation Advisory Committee (GMAC) which was prior located in the Department of Industry, Science and Tourism (for a history of GMAC, see chapter 2). An interim OGTR is already in place. The other two bodies will be the Gene Technology Community Consultative Group and the Gene

Technology Ethics Committee.[17] Regulation of GM foods, including labelling, is the responsibilty of the Australia New Zealand Food Authority (ANZFA) and Australia New Zealand Food Standards Council (ANZFSC). In Britain, three government departments, three 'strategic advisory bodies', and fourteen committees share responsibility under over-arching European Union regulations.[18] In the US the Department of Agriculture, the Food and Drugs Administration, the Environment Protection Agency, and the National Institute of Health have a stake in biotechnology regulation at the federal level. The individual States also have wide regulatory powers.

This pattern of managing biotechnological risks gives rise to two contrasting criticisms. On one side, the industry fears over-regulation. The Australian Biotechnology Association has complained that regulatory complexities 'can cause difficulties and delays and increase costs'.[19] Such complaints have become more muted as industry groups have grasped the extent of their public relations problem. On the other side are fears that current regulatory regimes cannot grapple with the broader and long-term risks of biotechnology. In an echo of Beck's thesis, the (British) Nuffield Council on Bioethics expressed concern that in case-by-case regulation 'no one is considering the long term implications of releasing [genetically altered] plants into the environment'. The council's secretary claimed that this concern is shared by many regulators.[20]

While there is presently little evidence of 'neo-liberal' risk management in biotechnology, it will surely develop. In the area of GM foods, as labelling becomes entrenched in Australia as elsewhere, a degree of responsibility for the management of risk will be shifted to the consumer in the neo-liberal manner. For example, it has been argued that some GM products might include traces of material that can cause an allergic reaction in some people. To adopt labelling, rather than prohibition, as the solution to this difficulty is to imply that responsibility is then passed to the susceptible consumer, who must diligently study the labels of all food products before she or he buys them. Again, if the genetic screening of human populations becomes established (as Gesche suggests in chapter 6), it is easy to imagine that parents will be made responsible for decisions affecting the genetic health of their offspring. This possibility is canvassed in Wilkie's dystopia where parents agonise that they cannot afford to buy the top-of-the-range IQ for their child from 'Genes R Us'.[21]

Over recent years the management of biotechnological risks has been made much more difficult by public distrust of experts and governments. In the most obvious example, since the BSE disaster in Britain (and now more widely in Europe) expert or governmental assurances that food is safe carry no weight with affected publics. Even without BSE, there is evidence from the US that attempts to reassure publics about biotechnology by putting regulations in place generally fail: government statements say 'biotechnology poses no special risks', while government actions in establishing special regulations say 'biotechnology poses special risks'.[22] This returns us to the point that rhetorical struggles over the meaning of biotechnology and the effectiveness of its regulation are a crucial dimension of the management of biotechnology risks. At issue is not the quantifiable risk of a specific adverse outcome, but rather a general public unease at the cultural riskiness of biotechnology, coupled with general distrust of experts and governments. This rhetorical, ritualised dimension of risk management does not align neatly with either the 'organised' or 'neo-liberal' regimes and requires attention in its own right.

Rhetoric, ritual & reassurance

We have seen that biotechnology is 'culturally risky' because it breaches boundaries that have long ordered Western culture. This riskiness could be managed if the boundaries could be redrawn, or if biotechnology could be shown not to breach them. The rhetorical battle over the cultural riskiness of biotechnology is fought along two main axes, one running between 'natural' and 'unnatural', the other between 'old' and 'new'. If 'new' and 'unnatural' are both risky, it is important to its proponents that biotechnology should not be seen as having both characteristics at once. The ideal, but implausible, strategy would be to position biotechnology as both 'old' and 'natural'. Failing that, it might do to have its 'unnaturalness' tempered by age, or its novelty tempered by 'naturalness'.

This rhetorical task is not something that is wholly external and alien to biotechnology. A number of commentators have noted that the 'new biology' frequently accounts for itself through metaphors of 'information' and 'codes'[23]: life becomes something that we can 'read'. These metaphors shape the way that biotechnology represents its capacity to control nature: control becomes a matter of 'rewriting' life in ways that better suit human purposes. Here is a powerful way of managing the cultural riskiness of biotechnology. If nature can be shown to be deficient—producing plants that are too sensitive to environmental stresses, or too easily attacked by parasites—then the re-writing of natural codes is a way of rectifying 'genetic deficiencies' in crops. Biotechnology offers modest improvements on an occasionally deficient nature, giving a helping hand to evolutionary processes.

The 'rewriting of life' is a strong and prevalent version of the claim that biotechnology is 'natural'. That biotechnology is merely an extension of nature, and consequently can be expected to be perfectly safe, forms one of main planks of many campaigns by Monsanto and other biotechnology companies. The assertion that there is a 'substantial equivalence' between GM foods and their non-GM equivalents has been widely used as an argument against the labelling of GM foods. Whatever its other (much disputed) merits, this argument ignores the question of *cultural* equivalence, of whether GM and non-GM foods are 'equivalent' in the classificatory schemes that members of the public use to distinguish between the 'natural' and the 'artificial'. It is notable that the lay panel of the recent Australian 'Consensus Conference' on GM foods firmly rejected the 'substantial equivalence' argument.

There are many variants of the 'extension of nature' move. The Australian Biotechnology Association argues that 'evolution itself results from successful mutations that occur randomly in nature. The new techniques of biotechnology simply increase the rate and precision of such changes'.[24] A US FDA-sponsored 'bST Fact Sheet' from Cornell University handles the issue with some finesse. It is stated that bST is 'normally' produced in the pituitary gland of dairy cows; it is one of 'a group of hormones produced naturally in the cow'. Nothing is mentioned about the method of production of r-bST, which takes place in the industrial laboratory rather than in the cow, and the acronym r-bST is used only twice, the favoured term being 'supplemental bST'.[25] The principle is simple: avoid any direct reference to obviously 'artificial' processes and choose words that convey 'naturalness'. 'Supplementary' is softer than 'recombinant', just as 'genetic modification' is softer than 'genetic engineering'.

The European Federation of Biotechnology (EFB) linked the issue of 'naturalness' with that of 'novelty'. Although the EFB acknowledged that people doubt the safety of GM products because genetic modification is perceived as an 'unnatural' process, the EFB claims that conventional science does not acknowledge any distinction between 'natural' and 'genetically modified'. To show how the lay distinction is inappropriate, the EFB points out that: 'almost all common foods in our diet come from the breeding, hybridisation, and selection of plants, animals, and micro-organisms over many centuries. These are also genetic techniques which could therefore be regarded as "artificial"'. The EFB emphasises the long history of 'biotechnology' by using different adjectives—but the same noun (biotechnology) for its 'traditional' (wine, cheese, and beer-making) and 'modern' (r-DNA and cell fusion) forms.[26]

This view of biotechnology as the up-to-date version of the wholesome cottage industries of the past is endorsed by the Australian Biotechnology Association, for whom the main difference between 'modern biotechnology' and 'traditional methods' in stock breeding is that the former produces 'hardier and more productive stock more quickly' than the latter. The same connection and contrast between the traditional and the modern can be found among the boosterish stories that have formed the staple diet of Australian press coverage of biotechnology. The link with tradition serves to ease any anxieties about the cultural riskiness of 'modern' biotechnology, while the break with tradition emphasises the superior efficiency of the modern. In a classic example of this technque, a *Sydney Morning Herald* story reported on a project seeking to establish a sheep-milk cheese industry at Cowra and using genetic-mapping techniques to select stock: it linked the product to the Rocquefort cheese mould, and its proponents to Julius Caesar and Charlemagne.[27]

That there must be a contrast, as well as a link, between the modern and the traditional is critical to the promotion of biotechnology. The modern must be like the traditional, only better. Another *Sydney Morning Herald* story reported an Australian robot—Vitron 501—that can clone three million trees a year.[28] It will accelerate the environmentally desirable shift from logging old growth forests to plantations, produce export income, and replace 'tedious' manual work that has a high risk of repetitive strain injury (RSI). Seedlings for cloning are identified through 'DNA fingerprinting' that allows selection after eight weeks rather than the traditional 30 years.

To summarise, the cultural riskiness of biotechnology is managed 'ritualistically' through 're-assurance regulations' and through a series of rhetorical devices that neutralise one or both of its dangerous characteristics—novelty and artificiality. The vigorous promotion of the benefits of biotechnology is also a tool for repositioning its riskiness: if there are a few risks, they are surely worth running. Biotechnology will feed the hungry, cure the sick, restore the environment, and reinvigorate the economy. Those who fail to see this are either ignorant or mistaken.

And the critics who would stem the biotechnological tide are viewed as placing in jeopardy human progress.

Will biotechnology win the rhetorical war?

Boosterism about biotechnology fits well with the dominant discourses on science in advanced societies, with their upbeat emphases on new discoveries, new medical treatments, and new technologies (see White in this volume). By contrast, critics of biotechnology appear to have at their disposal only the inferior resources of subordinate and marginalised discourses that articulate anxieties about environmental catastrophes, lurking threats to health and technologies that run out of control. On this basis, and on the evidence of the skill with which the cultural riskiness of biotechnology is neutralised, it may seem that the war over the cultural meaning of biotechnology has been won by its proponents.

However, such a judgment would be premature. Over recent years, the previously subordinate discourses have been given a major boost by the widespread collapse of trust in experts and governments noted earlier. Consumer resistance to GM foods has forced the hand of retailers. Burger King and McDonalds in Germany and increasing numbers of food producers in the US, as well as British supermarket chains (mentioned above), have announced a 'GM free' policy. Experts in the management of biotechnology risks have noted these developments, and have begun to give a greater emphasis to the need to re-build public trust in biotechnology. Governments, too, have begun to realise that their strategies of boosterism for biotechnology backed up by expert reassurances on safety are not working.

In Australia, a 'consensus conference' on gene technology in the food chain was held in early 1999, bringing together public or 'lay' representatives, experts, and government representatives. The report of the 'lay panel' expressed strong concerns about matters such as food labelling, 'open' assessment procedures, the need for public information and debate, and the power of transnational agribusiness to potentially control food production through GM technologies. The clearest impacts of these concerns have been the inclusion of lay representatives on GTTAC (the replacement for GMAC) and the establishment of 'Biotechnology Australia'. This inter-departmental unit, established in 1999, co-ordinates non-regulatory biotechnology issues for the Commonwealth government, and 'seeks to provide balanced and factual information on biotechnology to the Australian community'.[29] The British government has also taken steps to be more consultative. It has established a cabinet office 'Genetic Modification Issues' website that aims to explain 'key facts' and provide an 'overview of government policy'. The recent decision of the Australia New Zealand Food Authority to require labelling of GM foods should be understood against this background.

Whether efforts to rebuild trust through the reform of regulatory frameworks and the provision of 'information' can succeed remains to be see. In the present climate, regulators face a paradox. If they do not discover, or do not publicise, breaches of regulations covering GM organisms they will lose credibility. But if they do discover and publicise such breaches, they further erode public confidence in biotechnology. In the US, the recent scandal surrounding 'StarLink' corn managed to do both. StarLink, developed by Aventis, was not approved for human consumption but was discovered to be present in taco shells (see also chapter 9). It is thought that some nine million bushels of StarLink may have entered stockpiles of corn destined for human foods. Aventis ceased production of StarLink in September 2000, and have been attempting to trace the missing corn. The additional embarrassment in this case is that the discovery of StarLink in human food was made by consumer groups, not government regulators. On a much less spectacular scale, the IOGTR in Australia has produced critical audits of field trials of GM crops by Aventis and Monsanto that are alleged to have breached the conditions imposed by GMAC.[30] Predictably, the effect of publicity in these cases has not been to enhance trust in the effectiveness of the regulatory framework, but to decrease trust in trials of GM crops (see also chapter 2).

In conclusion, profound but unfocussed public anxieties about GM crops and foods arise within our contemporary 'risk societies' and are linked to biotechnology's cultural riskiness. Anxieties can be neutralised in ritualised practices and rhetorics, but these will be effective only if they and their sources are trusted. But those reassuring practices and rhetorics are increasingly *distrusted*, partly because they are recognised as aiming only to reassure. It is the toxic combination of cultural riskiness and generalised distrust that has forced the promoters of biotechnology increasingly onto the back foot over recent years. There is now a real prospect that consumer resistance will wipe out large sectors of the biotechnology industry in advanced societies. If governments wish to avoid this outcome, they must take a long-term view that gives absolute priority to rebuilding trust. This will require them to align themselves more directly with the positions taken by widely trusted consumer and environmental groups and to distance themselves much further from the agendas of the biotechnology industry.

However, such a strategy places governments in a difficult dilemma, because a dramatic re-ordering of priorities would itself drastically slow down the pace of development of biotechnology. The risk calculation that governments must make is one that weighs the certainty of much slower development in biotechnology on one side against the possibility of a complete collapse of public confidence that would devastate the industry on the other. It will be fascinating to observe how governments in Australia and around the world settle this calculation over the next few years.

Part 3

Molecular farming, biopiracy & campaigning

9

Gene technology, agri-food industries & consumers

GEOFFREY LAWRENCE, JANET NORTON AND FRANK VANCLAY

Scientists, government officials, and rural producers consider biotechnologies to be the latest in a long line of 'high-tech' inventions that will bring great advantage to Australian agriculture and food manufacturing.

In previous decades, Australian farmers have utilised the latest tractors, headers, ploughs, fertilisers, pesticides—and, of course, improved seed varieties and animal breeds—to attempt to reap profits and to remain internationally competitive. Similarly, food-processing companies have continually sought ways to enhance their market share by developing new products, and by reducing costs through innovative technologies.

For conventional farmers, new inputs allow for improvements in labour efficiency and for increases in plant and animal productivity. Improved plants may produce more grain, use less fertiliser, be harvested in a shorter time, or resist plant pathogens. Improved stock may convert pasture grasses more efficiently, tolerate ticks or drought, reach slaughter age more quickly, or better suit consumer markets. Conventional breeding programmes have enabled scientists to make small but important improvements to crops, pastures, and animals. But greater improvements are deemed to be necessary in the competitive national and international markets in which farmers operate. Farmers face circumstances often involving overproduction (which forces commodity prices down), so-called 'unfair' barriers to free trade such as tariffs and import restrictions, subsidisation of farmers by overseas governments which undercuts prices received by Australian farmers, and the overall tendency for farm input costs to rise faster than the rate of returns for farm output. Their challenge is to produce more, and more efficiently, and to do so in ways which enhance environmental sustainability.

The food industry faces many of the same pressures as the farmers. It has to compete with overseas products often 'dumped' on international markets, and must find means of ensuring that foods exported remain appealing to consumers in distant locations. This means preventing spoilage, increasing shelf life, and enhancing flavour—all within a competitive pricing regime. It is here that genetically engineered or recombinant-DNA (r-DNA) plants, animals, and manufactured foodstuffs are being touted as revolutionary new ways to attain efficiency gains in farming and in food processing *simultaneously* with improvements in environmental 'security'.

Agricultural biotechnology in Australia

Leaving aside 'basic' and human health research areas, the application of biotechnology to agriculture and the food industry (or agro-biotechnological research) fits within seven broad categories: animal health; resistance of plants and animals to diseases; pests and environmental extremes; the production efficiency of crops, pastures and animals; production efficiency in food processing; the quality of food and agricultural products; development of new products and processes (such as novel foods and biological pesticides); and the environment.

Recombinant-DNA research currently conducted both in Australia and abroad includes the cloning and breeding of animals (for example, transgenic animals, growth hormones); vaccine production for the sheep, beef and pig industries; the development of insect-resistant plants through the insertion of *Bacillus thuringiensis* (Bt) genes; the improvement of cellulose digestion in ruminants; the use of hormones to produce low fat meat (particularly in pigs); and gene-modification of plants and animals to allow them to produce complex proteins for use by the pharmaceutical industry.

Some projects have special relevance for Australia. CSIRO scientists are using r-DNA methods to alter forage plants so that they might produce more cysteine—a scarce amino acid. Sheep ingesting the gene-altered plant species will be able to produce more wool. Experiments are also being conducted to improve the ability of ruminants to digest cellulose. Given Australia's arid landscape, it is hoped that if digestibility of forage increases, so will the output of animals in marginal (as well as in the more fertile) regions.[1]

In a similar experiment, scientists are attempting to alter rumen bacteria so they can detoxify the plant poison, fluroacetate. If the research has a positive outcome, graziers in northern Queensland will overcome current stock-loss problems. CSIRO scientists have also successfully transferred a gene, from the common French bean into the field pea, which produces a protein causing starvation in insects. When insects begin to eat the pea crop they 'feel' full, but actually starve to death. Plant geneticists are also attempting to develop sugar cane plants that can tolerate frost, flowers which remain fresh for longer periods, salad vegetables which do not 'brown', bananas which contain doses of vaccines which prevent gastro-diseases in children, and novel plants including the blue rose. In September 2000, Monsanto gained approval in Australia for the commercial release of Roundup Ready cotton. This cotton plant contains a gene from a soil bacterium that yields the plant tolerance to glyphosate.[2] The list seems endless.[3] Mitchell Hooke of the Australian Food and Grocery Council has argued:

> The reality is, that there is no choice in embracing this new technology. People cannot, indeed, should not, attempt to divorce themselves from it. It is simply neither desirable [n]or possible. Unless new gene technologies are embraced and unless investment is made in its development and commercialisation, there is a real risk of Australia losing out, making farmers and food producers subservient to those holding the intellectual property rights to genetically modified material.[4]

For many scientists, the commercial imperatives are only part of the reason for pursing a biotech future. Recombinant-DNA technology is viewed as having major advantages for the environment. Research includes the production of biological or 'natural' pesticides (such as Bt) which reduce the need for the application of increasingly potent and synthetic agrochemicals; engineering plants and animals which need less water—viewed as highly desirable in a land where water is a scarce resource; and controlling pests like foxes and rabbits through induced sterility.[5]

In the food industry, the conversion of agricultural products, by-products, and substances currently wasted in food processing into value-added materials is a major thrust of research. Recombinant-DNA techniques are being employed to replace chemical additives in foods, and to enhance colour, texture, and other properties. Activity to date has been in improving starter cultures in dairy products, making cheese production more efficient through r-DNA derived rennet, developing 'natural' preservatives such as bacteriocins (which eliminate specific spoilage organisms), developing quality assurance tools, and improving the nutritional value of foodstuffs. Nationally, one of the main aims of public research is to discover ways of using agricultural surpluses via the conversion of bulk products into new substances.[6]

These examples indicate the general approach of Australian biotechnology research, and highlight the typical output-boosting, production-efficiency focus of much of the work. By increasing plant and animal resistance to disease, improving plant and animal efficiency and their capacity to survive in arid Australian conditions, research will allow producers to continue along the conventional 'productivist' path of intensive agriculture.[7] The concern is that, rather than providing an alternative to the current 'high tech' approach to rural production—one which has brought with it pollution and environmental destruction on a massive scale—it will reinforce the current trajectory toward specialisation, intensification and, more generally, toward ever-increasing surpluses and further pressure on the environment.[8]

Many ecologists, social scientists, and public-interest groups question the ability of companies such as Monsanto, which are major players in biotechnological development—but whose profits are currently based upon the sale of environmentally damaging chemicals—to provide long-term solutions to these problems (see Phelps' comments in chapter 12). It is argued that genetically engineered products will be tapered to the existing chemical strategies, thereby conforming to rather than challenging and overcoming agriculture's chemical 'fix'. As US sociologist Jack Kloppenburg once mused: 'I would like to think that—properly used—biotechnology can provide us with important tools to move towards a more sustainable agriculture. But can we trust Monsanto to get us there?'[9] Others posit, instead, the creation of 'superweeds' and 'superbugs' that have become resistant to the genetically engineered seeds which are packaged for the farmer, along with proprietary chemicals, and sold by agribusiness.

State support for gene technologies

With its history of publicly funded research in agriculture aimed at improving plant and animal productivity, Australia has reacted to the brave new world of genetic engineering by wholeheartedly embracing it from its inception in the early 1970s (as Hindmarsh details in chapter 2). It has paid for its scientists to go abroad to study the latest techniques, has conducted major studies and held important conferences, has identified biotechnology as one of Australia's main research foci, and has put in place a number of measures to stimulate the development of the bioindustry in Australia—including a National Biotechnology Program.

From the 1980s, the state has sought ways of promoting private capital investment and involvement in Australian agro-biotechnological development. In the CSIRO, the direction of research has moved from traditional means of breeding toward genetic engineering. Staff members were told that their promotional opportunities would improve if they developed industry linkages. This was in concert with two other developments. A commercial arm called SIROTECH was established to assist scientists to patent their inventions and to contract the development of products to outside agencies; and divisional chiefs were directed to obtain at least 30 per cent of their departmental funding from 'outside' (generally private) sources. Today, the commercial linkages of the CSIRO remain an essential part of the Australian government's vision for biotechnology which is to 'capture the benefits of biotechnology for the Australian community, industry, and the environment'.[10] It currently has a five-year alliance with the transnational Aventis and trains their officers in the so-called 'Aventis Academy' in Canberra. Aventis is the nation's largest private holder of GMO licences, undertakes the greatest number of field trials, and plans to have a farmers plant over a million hectares of its genetically modified canola by 2005.[11]

The apparent urgency by government and industry to further the 'productivity push' in Australian agriculture relates to Australia's balance of payments and debt problems. Large sections of Australian farming are facing severe economic and social problems. Strategic technologies such as biotechnologies are viewed as essential to improve the economic position of our farmers and to address Australia's trade imbalance.[12]

According to Scott Kinnear of the Organic Federation of Australia (OFA), the strong relationship between government and biotechnology firms resulted, in December 2000, in a Gene Technology Bill that has the potential to compromise public health and safety, place the environment at risk, and have a negative impact on farming.[13] It is argued that with a 'general release license' the Gene Technology Regulator will not be able to impose conditions to limit pollen flow unless there is a significant risk to public health and safety. Any conditions on GM plantings that are trade-related are outside the power of the legislation. In addition, farmers will not know where GMO crops are being grown. Australia stands to have the weakest legislation in the world. It will not take into account economic impacts on organic or on other non-GMO producers. The OFA predicts widespread contamination of organic and non-GM crops if general release licenses are granted under this legislation. This is particularly important for organic producers who obtain premiums for marketing organic GMO-free produce. The cost of segregation and testing of organic and non-GM crops will also be unfairly allocated to the producers of these crops. And, if they do suffer contamination, they are expected to identify the exact location of the source of that contamination. Otherwise, any claim for damages is unlikely to succeed.

Gene technology & Australia's integration into the global food order

A most enthusiastic voice for biodevelopment as a key avenue for Australia's integration into the Asia–Pacific region has been that of Australia's peak establishment farming organisation—the National Farmers' Federation (NFF). Unlike its Canadian cousin, the National Farmers' Union, which has recently demanded a moratorium on the production, distribution, importation, and sale of GM foods,[14] the NFF endorses the Australian government's biotechnology thrust, and agrees with the OECD's prediction that without biotechnology the rate of increase in agricultural productivity will soon fall. Moreover, according to the NFF: 'Trends in rural production already suggest that future differences between the output of developed and developing countries will be based on their respective level of biotechnology use.'[15] Biotechnology is thus viewed as being vital to expanding the sales of Australian primary products—and especially in the role of helping to feed a burgeoning Asian population.

Biotechnology Australia—the federal government agency responsible for supporting the development of biotechnology—considers that biotechnology is a 'strategic' technology in the processed food industry.

The potential worldwide biotechnology food market is estimated to be worth $US12 billion.[16] Complete endorsement for a biotechnology future for Australian foods has come from the corporate-linked Australian Food and Grocery Council (AFGC). A glance at the website's various 'fact sheets' and news releases indicate the extent of boosterism occurring. Gene technologies are considered as safe for consumers, as improving the environment (through, for example, Roundup Ready soy-beans), to be the key to feeding the world's expanding population, and to be the future for food manufacturing. Opportunities are seen to exist for food processing firms to form backward linkages to agriculture, and for chemical and biotechnology firms to become involved in food pro-cessing. Indeed, today, many of the largest and most prominent of the food manufacturing firms operating in Australia use modern biotech-nology, including genetically modified ingredients.[17]

Importantly, both the AFGC and governments at various levels believe that Australia can position itself as the 'clean' food nation through the incorporation of GM technology. Biotechnologies are seen to underpin Australia's advance into the unprocessed and processed foods sectors in the Asia–Pacific region. According to the federal gov-ernment's Bureau of Rural Resources:

> It is essential to have public acceptance of recombinant DNA technology if we are to move to have commercially viable industries in agriculture and in biotechnology. It must be stressed that it is not just the biotechnology industry *per se* and its export potential that is at stake but the survival and competitiveness of our major industries in agriculture, both plant and animals that are also at stake. Major sections of the rest of the world are moving into using genetic modification to improve the efficiency of agriculture and we cannot afford to be left behind with outdated technology.[18]

Here, then, is clear evidence of the endorsement of a biotechnology future for Australian agriculture and food industries. In fact, six devel-opments have converged to promote biotechnology—and particularly genetic engineering—as a key component for an agriculture and food industry of the new millennium. These comprise the need to boost farmers' productivity in a more highly competitive global trading arena; scientific advances in biotechnology; the state's demand that publicly funded science be more closely linked to private industry; the growth in

importance of large trading firms—especially in relation to their connections with Asia; new consumer markets in the Asia–Pacific region;
and public demands that farming systems become more sustainable and
that environmental degradation and pollution be overcome.

The question is: will the application of biotechnology fulfil current
expectations?

The uncertain promise

The issue of whether farmers will benefit from the new biotechnological inputs is of great importance. If agribusiness produces and
distributes the new genetically modified seeds and livestock, there is no
guarantee that rural producers will benefit. These inputs will, after all,
be the exclusive (private) property of the companies developing them.
Today, input prices are rising much faster than returns for farm goods:
costly biotechnological products might be expected to reinforce this tendency. Some question the likelihood of benefits flowing to farmers as a
group, and note the current polarisation in agriculture, claiming that an
agribusiness future will be an industrialised agriculture with fewer, more
wealthy, farms among a great many smaller, economically marginal,
producers unable to afford the new expensive technologies.[19] For
example, herbicide-tolerant plant species are viewed not as freeing agriculture from chemicals—one of the promises of biotechnology—but of
ensuring the dependence of farmers on proprietary brands of herbicides.
This has been referred to in colourful terms as 'bioserfdom'. If herbicides increase their presence, there is the real possibility of chemical
resistance occurring in weeds, and of increased chemical pollution in
waterways—again, the opposite of the biotechnological promise.
Indeed, the United States Department of Agriculture reported in 1999
that genetically engineered soy, corn, and cotton grown in some areas
of the United States neither gave significantly increased yields nor
reduced herbicide use.[20]

With respect to the claims that biotechnology will clean up the
environment and reduce the amount of land required for production, it
is necessary to question the possibility for either outcome in Australia.
While some biotechnologies may have the potential to eliminate
introduced feral pests (through biotechnologically induced sterility), to
clean up oil spills, or to reduce the use of some herbicides and
pesticides, the introduction of GMOs in the environment may well result
in the transfer of genes between species (including the imparting of

resistance in undesirable bacteria and plants), increase the risk of escape of pathogenic organisms, and disrupt and potentially damage biotic communities.[21] As one critic suggested 'engineered (species) cannot easily be both widely used and closely contained. There will inevitably be some risk of genetic pollution, and in this case it will be pollution by genetic material unprecedented in nature'.[22]

While increases in the production efficiency of 'enhanced' plants and animals may, in theory, allow for reductions in land use, it is more likely that farmers will seek to achieve greater volumes from existing lands—as a strategy to offset the cost-price squeeze through greater levels of output. In other words, the intensification of agricultural production will continue unhindered, thereby increasing pressures on Australia's fragile environment.[23] It may also see new transgenic crops and livestock developed for Australia's more marginal inland regions— potentially wreaking environmental havoc in those areas. Farmers themselves are not necessarily convinced of the safety of the new technology. A recent survey revealed that that 70 per cent of farmers considered GMOs to be only partially safe with concern expressed over potential unforeseen side effects.[24]

There is a growing list of concerns regarding release of GMOs. Three main issues are of particular note. The first is an ecological one: Australia is particularly vulnerable to 'foreign' or 'exotic' organisms and the potential extinction of indigenous species, including the incompatibility of GMO release with Australia's parallel desire for biodiversity and sustainable development. The second issue concerns the future of the organics industry in a world of biotechnology. The third relates to food and consumers: is the extension of biotechnology compatible with the trends in 'green consumerism' both in Australia and in food-importing nations? These are discussed below.

Ecological vulnerability and biodiversity

In terms of ecological vulnerability, problems caused by the past release of foreign species designed to improve the profit levels of farmers are poised to occur again. For example, when the South American cane toad (*Bufo marinus*) was introduced via Hawaii to control the cane beetle, it was not foreseen that it would impact on other species, or that it would expand its niche beyond the cane fields. It has now spread out of control in Queensland, the Northern Territory, and New South Wales, leading to the death of native animals that prey upon amphibians. For example, numbers of the red-bellied black snake has been severely reduced through eating cane toads. In turn, this has allowed the population of its natural enemy, the more venomous brown snake which do not eat frogs or toads, to increase in number. The cane toad is also out-competing native frogs. Insufficient research was conducted to establish the likely impact of the toad. The uncontrolled spread of the calicivirus following its accidental (some speculate, deliberate) release on the Australian mainland in 1996—while having an overall beneficial out-come of substantially reducing the wild rabbit population—confirms the inability of scientists to contain their field experiments. The ensuing illegal introduction by farmers of the calicivirus into neighbouring New Zealand is yet another example. Moreover, there is always the possibility of willful neglect. In July 1997 a scientist at the Queensland University of Technology was suspended for knowingly exposing colleagues to the risk of contracting Japanese encephalitis while he undertook unap-proved experiments on the deadly virus. Although these three cases are not related to genetic engineering, they nevertheless confirm the public's fears about the difficulties faced in keeping scientific and genetic exper-iments under control.

This begs the question: with strict guidelines and a stringent moni-toring system in place in Australia, are these sorts of 'accidents' likely to occur in relation to genetically engineered species? In fact, one already has. It occurred in South Australia in 1988 with the release of transgenic pigs (see also chapter 2). Without obtaining full permission, the firm Metrotec sent a batch of the pigs to Adelaide's abattoir for slaughter. While, in this case, there was virtually no potential for the escape of 'new' genes from the animals, it represented a major breach of government guidelines. The point is, with so many rural producers wanting immediate productivity gains, and with biotechnology compa-nies wanting a return on their substantial investments, there is inevitable pressure for the release of novel organisms. As Wills argues in chapter 3, scientists have only a limited understanding of the ecological com-plexities of existing systems, and cannot give any certainties that novel genes will not transfer to native species.[25]

The issue of foreign genes in the environment is linked strongly to concerns about biodiversity. If new, productive, genetically altered species are economically more beneficial for producers, they may be adopted *en masse*, reinforcing current patterns of mono-varietal production in Australia. Yet, despite short-term productivity gains, they might not provide long-term economic or ecological benefits. The narrowing of the genetic base for agriculture is inconsistent with Australia's avowed desire to embrace ecologically sustainable development (ESD). In its draft national strategy of 1992 the ESD steering committee stressed the need for both nature conservation and environmental protection as part of a broader strategy to protect diversity. In this report—and in the document of the Working Group on Sustainable Agriculture which appeared a year before—the complicated issue of biotechnology's role in relation to biodiversity was largely ignored.

It should be noted that various surveys in Australia indicate most rural producers and bioscientists believe strongly that genetically engineered plants will reduce pesticide use, are compatible with the nation's 'clean and green' image, and that they will contribute to a more sustainable agriculture. Those largely unconvinced are organic growers (see discussion below), dissenting scientists within biology and ecology fields, and consumers. Women, in particular, appear more concerned than men about environmental and social consequences.[26]

The organic farming and food industry

Organic agriculture is based upon the minimisation of the use of external inputs into the farming system, with particular stress upon the avoidance of synthetic fertilisers and pesticides. The use of GMOs in organic production is strictly forbidden—in Australia and overseas. To be labelled 'organic' a food must meet rigid standards throughout its production, handling, processing, and marketing stages, and then be certified by an approved body. Such adherence to regulation assures consumers that products with the organic label have certain desirable features setting them apart from conventionally grown foods. The strong worldwide growth in consumer demand for organic foods is associated with the 'clean and green' image of organics, reflecting the desire by consumers for more nutritious, more 'natural', and less chemically laden foods.[27] It also highlights the public's desire for more sustainable forms of agriculture, with a growing number of people identifying organics with healthy farming systems.[28]

In Australia, the value of organic production has doubled since 1995 and is currently worth $A250 million (about $80 million of which is exported). Worldwide, the industry is expanding at a rate of between 20 and 50 per cent, and is expected to be worth some $150 billion by 2004.[29] It has been predicted that over 30 per cent of Europeans will be eating organic foods within the next ten years, with the demand in most industrialised nations growing markedly. Not surprisingly, an increasing number of Australian and overseas farmers want to be part of this valuable new market.

In recognition of the economic benefits of organic production, some State governments (notably NSW and Queensland) have begun to support the growth of organic farming. But they are doing so with funding which, in relation to that going to conventional farming and genetic engineering, is miniscule (see chapter 12). They also remain largely deaf to the cry from organic producers that the release of genetic materials into the ecosystem will seriously compromise the integrity of, and damage sale prospects for, organic foods.

The debate about a GM versus an organic 'pathway' in agriculture is one of the most important occurring in Australia today. According to the pro-biotech writers such as the US-based Dennis Avery, organic production cannot sustain soil fertility, actually (via tillage methods) increases soil erosion, does not provide the 'balance' between production and pests claimed by proponents, and will be largely incapable of feeding the world's growing population.[30] According to Youngberg and his co-writers, there is a negative symbolism associated with organic farming which has not yet been removed. In the US, organic agriculture is viewed by many as a 'primitive, backward, nonproductive, unscientific technology suitable only for the nostalgic and disaffected back-to-the-landers of the 1970s'.[31] Here the agricultural power elite reinforces the view that it is the *real* science of modern, technologically based agriculture and its GMO-based approaches that will prove to be the most effective and enduring for the new millennium.

Not surprisingly, the organics industry rejects these views, expressing five main concerns in relation to the release of GMOs. The first is that the form of science upon which genetic engineering is based is reductionist and that, in both failing to recognise the complexities in nature and in dealing with symptoms rather than causes, it will inevitably lead to partial or inappropriate 'solutions' to the problems of food production and sustainability (see also chapters 3, 10, and 12 in this volume). The science of which genetic engineering is part is therefore not considered to be the basis for clean and green foods, or for an enhanced environment. The second concern is that GM pollen will pollute non-GM crops, rendering the latter incapable of being labelled 'organic'. The premiums now available to organic producers will therefore be eroded, or will disappear entirely—as might certain sections of the organic industry. The third concern relates to the use of the Bt gene in GM plants. Bt is used extensively by the organics industry as a natural pesticide. When it is built into plants a component of large-scale monocultures, insect populations will develop resistance, in the way they currently do to chemical pesticides (see chapter 10). If/when pests develop a tolerance to Bt the organic producers will be deprived of a proven, natural, way of fighting insect pests. Fourth, in relation to the strong consumer interest in organic foods, coupled with a growing overseas rejection of genetically altered foods, the release of GMOs into the environment is likely to destroy an exporting nation's 'clean and green' image. Here, countries like Australia and New Zealand would be particularly vulnerable if overseas consumers begin to reject foods from nations which have gone down the biotechnology path.[32] This is further explored, below.

Consumers and foods

Most agricultural products move along the food chain destined, as they are, for either human or animal consumption. Even the production of natural fibres such as cotton and wool results in by-products such as cotton seed oil and mutton. The issue of whether consumers will be prepared to purchase and eat the new gene-altered products is therefore important to agriculture. Increasingly more discerning consumers— concerned about food safety and nutrition—are eschewing chemically laden products, purchasing 'clean and green' or 'natural' products, and are demanding the labelling of any genetically modified (GM) (or so-called 'foreign') foods. Growing evidence from home and abroad shows that consumers are wary of genetically modified foods and that there is a strong and growing market for foods that can be guaranteed to be free of genetically modified ingredients.[33]

In Europe, consumers consider that GM food threatens the 'natural order', could be the source of global disaster, presents danger to future generations and is 'simply not necessary'.[34] This is consistent with previous results that show that support for genetic engineering is decreasing in Europe. Within Australia, Brazil, Canada, Japan, the United Kingdom, and the United States consumers were found to have a negative view of GM foods and considered that the risks of these foods outweighed the benefits. Further, consumers considered the issue related to these foods was health and safety rather than scientific progress. A detailed Australian study demonstrated that acceptance of genetically engineered products was dependent upon the type of gene transfer occurring, perceived benefits to the organism or the environment, and whether the product would be ingested.[35]

Even if biotechnology improved output and efficiency in Australian agriculture, consumer rejection of the foods so produced might undermine any gains. Increasing concern about the safety of food and the nature of additives has meant that issues relating to the overall acceptability of food have become prominent. The changes in food consumption patterns of modern societies, the survival of traditional culinary practices and beliefs within those societies, the emergence of professional nutritional discourse and health-oriented dietary advice, and the contradictions between them, combine to produce an atmosphere of uncertainty in the public mind. Within developed countries, food consumption has been assessed in terms of the risk of development of dietary-related diseases such as diabetes and coronary heart disease.[36] The nutritional value of food is generally assessed in terms of biology, physiology, biochemistry, and pathology with little emphasis placed on culture, society, behaviour, and the production processes of modern foods. This emphasis on the 'science of food' rather than societal influences highlights the present arguments about the introduction of genetically engineered organisms into the food chain.

Consumers have also expressed concerns over the ability of science and technology to forecast accurately the potential impacts of these organisms and to keep risks to a minimum. Past experience has resulted in a lack of trust in science practitioners, with consumers concerned about effects currently being felt by the overuse of herbicides and pesticides and non-renewable resources, as well as the greenhouse effect and depletion of the ozone layer, and pollution. The safety of the new foods are of concern to consumers who consider that rigorous testing is required prior to the commercial release of the new agricultural products. The unknown effects on the environment and long- and short-term health effects are of concern to consumers.[37] The uncertainty regarding some products has led to the closing of a Michigan food plant run by the Kellogg food company. Unable to buy corn that is not contaminated by the genetically engineered corn variety Starlink, Kellogg has been forced to close the plant. Starlink was not approved for human consumption because of concern that it may cause allergic reactions. Pressured by the US Agriculture Department (USDA), Aventis, the developer of Starlink, is attempting to buy back the crop by offering a 25 cent per bushel premium to farmers (see also chapters 8 and 12).

Within Australia, consumer acceptance of genetically engineered products has been found to be dependent on a person's concern about food safety rather than on knowledge or understanding of the technology involved. Women exhibited a greater wariness of the technology, and expressed reluctance to both buy and eat the products. Even when genetic engineering is acceptable, consumers are demanding products to be labeled as such.[38] The Australian government recently approved the labelling of GM foods with some exemptions such as food prepared at point of sale including restaurant-prepared food and takeaways.[39] With women as primary shoppers in most households, and with correct labelling in place, the success or otherwise of genetically engineered foods will ultimately be determined at the supermarket checkout.

Yet, for the pro-biotechnology Australian Food and Grocery Council, consumers are viewed as being out of touch with the benefits of biotechnology. While some people have 'real moral or religious concerns', much of the public's concern about food biotechnologies is seen as arising from a 'lack of understanding and knowledge'. Public education is seen as 'vital' for 'successful integration of genetically modified foods and food products into everyday life'. Scientists are endorsed and praised by the Council for ensuring that 'experimentations in recombinant-DNA techniques' are 'controlled and safe'.[40]

Despite the reassurances of the AFGC, consumers consider that there should be rigorous testing of products prior to release. Consumers are concerned that there should be long term testing of products to ensure that GMOs pose no danger to health. The testing required by consumers is not necessarily that which is already being performed (see also chapters 2 and 12 for a critique of Australia's regulatory system). In the words of nutritionist, Rosemary Stanton:

> I do believe there should be more of the right sort of testing before things are put on the market commercially. It's true that there has been a lot of testing already, but it has focused on identifying whether the proteins, fats, minerals and vitamins in modified foods are substantially the same as in unmodified food.[41]

Testing would be used to determine if the organism contained any unexpected harmful, toxic, carcinogenic, mutagenic or allergenic substances. Without testing of this type, consumers will remain sceptical of the technology, and markets will close to them. Increasingly negative consumer feedback has resulted in companies such as Sainsbury's in the United Kingdom, and Gerber to announce that their products will be free of

GMOs.[42] Australia must decide if it has more to gain by capitalising on markets that want food that can be guaranteed GM-free.

Conclusion

To improve Australia's competitiveness in the global marketplace, Australian governments, farmer groups, and food companies have endorsed agro-biotechnologies as the best option for productivity enhancement. They have been endorsed as the 'future' for farming on the basis of Australia's traditional comparative advantage in the production of farm products, the potential of sales growth in the Asian–Pacific market, and opportunities for the commercialisation of intellectual property. Biotechnologies are viewed as improving productivity at the same time as they improve agricultural sustainability. But can they fulfill this promise? Scientists' claims that the genetic engineering path will be one leading to increased output and efficiency are being challenged by recent research. Even if both were achieved, Australia might face the prospect of having its exported foods rejected by increasingly discerning consumers. We might have greatly increased production, but be left with unsaleable products, further eroding an already depressed farming sector. Australian farmers are, from our interviews and discussions, becoming increasingly aware of such a possibility.

Consumers are aware of the benefits of genetic engineering but they are also aware of the risks. The possible global implications of these risks, and the unknown effects that could be felt by future generations cause them much concern. Consumers evaluate these possible risks in the light of other, older, technological advances that have promised great benefit but which are now viewed as having caused health problems and environmental damage. As a result, the public now views the promises of science and technology with a degree of cynicism, and is not surprised when the side-effects of technology are exposed. This does not mean that people are anti-science but rather that they increasingly wish to be allowed to make their own choices after being given appropriate information.

Government, scientists, and food processors must recognise that consumers' concerns over genetic engineering are valid, and should be addressed rather than dismissed as modern-day luddism or an outcome of pure ignorance. Scientists must negotiate the future amid complex power relations. In the past, science was unquestionably hegemonic.

Now, consumer and other groups are lobbying to ensure that science works within boundaries set by a somewhat cynical public.

It would be prudent for Australians to give considered thought to the sort of food production system they want in this new millennium. If they have concerns about gene-technology, they will have to act quickly— and politically—to counter the already purposeful industry and government-endorsed strategy to pursue a genetically engineered future for the nation's agricultural and food industries.

10

Wearing out our genes? the case of transgenic cotton

ANNA SALLEH

In August 1995, the head of CSIRO's Division of Plant Industry, Jim Peacock, suggested to a scientific conference in Newcastle how the problem of world hunger might be solved. In a keynote address to the Australian and New Zealand Association for the Advancement of Science, he said that gene technology could bring the 'doubly green revolution' needed for agriculture to feed and clothe the world's growing population in an environmentally sustainable way.[1]

A year later, Australia's first genetically engineered broadacre crop was sown—30 000 hectares of 'Bt cotton' designed to kill its own pests. The new technology promised to eliminate up to 90 per cent of the insecticide sprays normally needed.[2] Saving the planet aside, however, cotton farmers had other, more immediate, concerns on their minds.

Cotton is one of the most heavily sprayed crops in agriculture. If the new genetically engineered variety could reduce spraying, it would help stem the tide of accusations that cotton growing was poisoning people and their environment. Not only that, but the technology could provide a timely solution when the arsenal of effective chemical weapons against crop pests was losing its effectiveness.

Some have described Bt cotton as one of the most successful applications of agricultural biotechnology research so far, and a major step towards a more environmentally friendly cotton industry. Others claim it is just another in a long line of quick fixes that characterises a crisis management approach to pests. They insist that it will be 'smart farming' rather than 'smart crops' that will deliver the much sought after bounty of sustainable agriculture.

This chapter assesses Bt cotton's potential for being a 'clean and green' technology. As such, it will explore one of the major risks it brings—that of 'wearing out' our valuable Bt genes.[3]

Cotton's pesticide treadmill

Monoculture cotton is more prone than many other crops to infestation by weeds and a variety of insect pests. The most formidable pests of cotton in Australia are the heliothis caterpillars of *Helicoverpa armigera* and *Helicoverpa punctigera*. Secondary pests include green mirids, mites, aphids, and thrips. Australian cotton growers spend up to $100 million a year on pesticides,[4] over one-eighth of Australia's total expenditure on crop protection products.[5]

The average crop is sprayed between six and sixteen times per season. Eighty per cent of these sprayings are against heliothis.[6] When pests adapt by developing resistance, however, many more sprays are used. For example, in 1974, just before the cotton industry at Kununurra in Western Australia was wiped out due to resistance, farmers were spraying 40 times a year in an attempt to control insect pests.[7] *Helicoverpa armigera* has developed resistance to most of the chemicals that in the past provided effective control.

Resistance aside, cotton growers also have a severe image problem. They are often portrayed as environmental pariahs. A relatively small percentage of cotton pesticides applied by air reach the target insects. This means that everything from humans and fish to other crops and export cattle may be exposed to these chemicals as well. While the effect of pesticides on the environment and human health is the subject of intense controversy, the negative impact of residues on the highly valuable export beef market is undisputed.

As the insecticide stockpile dwindles—either due to resistance or concerns over implications for the environment, human health or trade—the monoculture cotton industry is in crisis, and is desperate for a new weapon against heliothis. Over the last decade, biotechnology promised a dazzling solution in the form of transgenic Bt cotton.

Benign Bt: the perfect candidate for the first smart crop

Bt toxin is produced in nature by a soil bacterium called *Bacillus thuringiensis* (Bt). Of 20 000 different strains, the variety *kurstaki* has been of most use to agriculture. Sprays containing Bt *kurstaki* have been used worldwide to kill Lepidoptera caterpillar (including the diamond-back moth and heliothis pests) in vegetables since the 1920s. Bt is extremely valuable to organic farmers as it is both completely 'natural' in origin, and is quite specific to a small range of insects, while other insects—some useful as predators or parasites in biological control—are left unharmed.

Two years after recombinant-DNA technology was first applied to agriculture in 1981, Bt was identified as a perfect candidate to exploit. Yet, when the Cry IA(c) gene from Bt *kurstaki* was first transferred into cotton it did not produce very high levels of the toxin. In 1988, molecular biologists from bioscience company Monsanto 'edited' the gene to make it more 'plant-like'. This 'construct'—or designer gene, now patented by Monsanto—performed much better than the original.[8] It was transferred by CSIRO into local varieties of cotton, more suited to Australian growing conditions, in order to produce an economically viable crop.

The advantage of transgenic Bt cotton over Bt sprays were said to be many. Whereas Bt toxin in sprays was exposed to UV light which caused it to break down and become ineffective, Bt toxin in transgenic crops was protected inside the plant cells, where it remained active for longer. Another advantage was that the toxin was produced throughout the whole plant and at all times. This would ensure that caterpillars, no matter where they were on the plant, would be exposed to the toxin at first bite. Sprays, in contrast, often missed parts of the plant (such as the underside of leaves), or were only applied after the pests had already done some economic damage. With the advantages of transgenics, however, came a challenge—that of keeping pests from becoming resistant to the Bt toxin in the transgenic cotton, just as they had to all other insecticides used against them.

Resistance to Bt

In every insect population, some individuals possess genes which give them greater tolerance to a particular insecticide. While the chemical will kill most of the population, these 'resistant' individuals survive and, over time, breed up in numbers. When this happens, farmers tend to use higher and more frequent doses of insecticide—each spray inadvertently selecting for the survival of even more resistant insects. Eventually, such a high proportion of the population is resistant that the insecticide becomes effectively useless.

Already, in areas such as Malaysia and Hawaii, following a number of years of intensive use of Bt on vegetable crops, the diamondback moth has developed resistance to the same variety of Bt now being used in transgenic cotton. There are signs that similar resistance may already be developing in Queensland. Bt was used sparingly as a spray for over thirty years, and it was only when farmers had lost almost all of the synthetic insecticides to resistance, and concerns over pesticide residues increased, that Bt sprays began to be used more frequently.[9] Indeed, the use of Bt sprays in cotton has increased twenty-fold in the last five years.[10]

Today, because Bt is so valuable to organic and biological farmers it is used sparingly as an emergency control should other methods fail, and so that resistance has less chance of developing. The approach taken by organic producers—of minimising reliance on any one particular 'tool'—is advocated by leading scientists in the area of pest management research. This approach is referred to as Integrated Pest Management (IPM), and also happens to be the catch-cry of those who promote transgenic Bt crops.

The 'perfect' promise of integrated pest management

IPM—and the more recent Integrated Crop Management—spreads the selection pressure for resistance around by employing a number of pest management tools. These can include chemicals, predators, parasites, pathogens, pheromones, baits and attractants, resistant crop varieties, physical and cultural controls (for example, legislation), and biotechnology. Some, such as biological controls, are less likely to lead to resistance than others. Given the right balance of tools, some argue, it may be possible to avoid resistance developing altogether.

Biological controls, which many see as the mainstay of good IPM, require a healthy and diverse population of insects in the field which is not generally possible in monoculture cotton regularly sprayed with broad spectrum chemicals. While such chemicals act to turn the field into a 'biological desert', the specificity of Bt toxin to Lepidoptera caterpillars makes it very compatible with biological controls and thus IPM.

Transgenic Bt cotton, it was said, could provide a new 'platform' for IPM. This vision was held despite the diamondback moth experience, and despite concerns that the constant expression of Bt toxin in transgenic cotton posed even stronger selection pressure for resistance than Bt sprays.

Around the world, IPM practitioners have been applying the 'evolutionary brakes' and removing selection pressure wherever pests have shown signs of resistance to Bt. UK-based IPM researcher Robert Verkerk argues that transgenic cotton's built-in Bt toxin makes it hard to do this. 'You can't very well ask the farmers to pull out the crop half way through a season,' he says.[11]

Bt cotton researchers fully understood this flipside to the new technology which is why they developed a resistance management strategy that some hoped would stave off resistance for 'many decades'. But will this be so?

Resistance management for Bt cotton: bugs in the system

No matter how effective an insecticide, a certain number of tolerant individuals will always survive. The higher the dose of insecticide, the fewer the survivors, but the 'tougher' they will be. In any resistance management strategy it is crucial to deal with these lest they breed up into a whole army of immune 'monster bugs'.

When it came to Bt cotton, researchers reasoned that a suitable field of Bt-free plants located near the transgenic crop could provide a safe haven (or refuge) for heliothis to breed up in, without hindrance from Bt. The insects not exposed to Bt would be more 'susceptible' to the toxin, and by mating with any survivors from the transgenic crop could dilute out the undesirable trait of resistance in the next generation. These refuges were central to the resistance management plan.

Researchers estimated a refuge size of 10 to 20 per cent of the crop to produce enough susceptible insects to mate with resistant survivors in the transgenic crop. They also recommended a range of other controls for farmers using Bt cotton—some of these were already in strategies to manage resistance to previous insecticides. These included, first, cultivation (ploughing) of soil after harvesting to a depth of ten centimetres, to destroy any highly resistant pupae and their burrows; and, second, monitoring with a view to spraying (with their usual insecticides) should a 'threshold' number of heliothis caterpillars be detected.[12]

The refuge strategy was designed to work most effectively when a *consistent and high* expression of the insecticidal protein occurred *throughout* the plant. The idea was that the dose of Bt produced by the transgenic crop would be high enough to kill virtually all heliothis, leaving only a handful of highly tolerant individuals which could be easily swamped by an army of 'susceptibles' bred up in the refuge. This would minimise damage and reduce the chance of a resistant population developing. This was called a 'high dose' strategy.[13]

Yet, far from being the perfect vision of precision described in public relations brochures, transferring genes across nature's boundaries proved instead to be a very hit-and-miss affair. Consequently, Monsanto 'took what they could get'.[14] While the dose was high enough for most of the US heliothis pests (for which Bt cotton was initially developed),[15] Australian species were genetically more tolerant to Bt which meant more would survive leading to a high risk of resistance. Although researchers tried to compensate for this 'low dose' by recommending larger refuges than those used in the US (to produce more susceptibles), they were still worried.[16]

Bt toxin is a protein and, in transgenic cotton, like other proteins, its level in the plant changes according to factors that affect the plant's growth. NSW Agriculture entomologist Neil Forrester assessed the dose produced would be 'just enough' when conditions were good, but the plants would 'not need much of a hiccup to fail' making resistance 'a real problem to manage'. The outcome was that a significant percentage of the Australian 1996–97 Bt crop had to be sprayed more than expected. Growing conditions were suspected of being partly responsible.[17]

Entomologist Rick Roush from Adelaide's Waite Institute, declared that the crop did not produce enough Bt toxin to be 'conducive to resistance management'.[18] Despite Monsanto claims to the contrary,[19] the expression of Bt toxin in Australian transgenic cotton under real world conditions was neither high nor consistent.

Queensland University entomologist Myron Zalucki suspected the 'high risk' strategy would probably not work. Apart from the 'low dose' of Bt, Zalucki predicted it would be almost impossible to ensure that the susceptible insects from the refuge were around at the same time and place to be able to mate with the resistant ones. If, instead, resistant mated with resistant, super-resistant individuals might be produced in the next generation. Moreover, he said, insects on the Bt cotton grow more slowly than those on a non-Bt crop.[20] Thus, even if flowering of the refuge crop could be synchronised with the Bt crop, susceptible and resistant insects might still not meet. Also, researchers were not sure how far moths would travel before mating. In 1996–97 it was not compulsory for refugia to be adjacent to the transgenic Bt crop. Given this, if farmers see refugia as less productive, they may—especially in hard economic times—choose to put them on marginal land possibly too far away from their Bt crops to be effective.

Since the refugia requirement is a condition of Monsanto's licence, it is the company's responsibility to ensure compliance. Even though compliance is legally binding, even normally apolitical agricultural researchers have expressed concern that this apparent self-regulation is inappropriate. Despite Australia's international reputation for good resistance management, farmers have not always followed the requirements of previous strategies. Those on tight rotational programs, for example, prefer to directly drill their wheat crop into cotton stubble rather than plough up the field first, so they may be reluctant to cultivate their cotton fields after harvest as required under the current plan. The strategy also conflicts with the desire for minimum tillage agriculture. While technology to make tillage more environmentally friendly is reported to be under development, the 1996–97 crop was grown without it.

Meanwhile, one of the other reported advantages of Bt cotton—that it requires less spraying—was also under challenge.

To what extent will Bt cotton eliminate spraying?

Long before commercial Bt cotton was planted, Australian researchers knew that when the plant stopped growing towards the end of the season, Bt toxin production would drop off, and the cotton plants would be vulnerable to heliothis attack. While entomologists were worried about the increased resistance this would bring, growers who had massive overheads to cover would be more concerned about an uneconomic drop in yield. If they stopped the normal spray regime for heliothis they might also have problems with secondary pests, unaffected by Bt. It was all very well to have a long-term vision of a more diversified approach to pest management free from broad spectrum sprays, but in the short-term, researchers anticipated that farmers would spray Bt cotton two or three times in the season (even if these sprays were likely to be broad spectrum, and thereby defeat the selectivity of Bt cotton by killing beneficial insects).

What researchers did not anticipate was the degree of unpredictability of the transgenic crop in the real world situation. While early estimates were that there would be a 90 per cent drop in sprays, for the

1996–97 Bt crop as a whole there was only a 52 per cent decrease. The decrease was not evenly spread among producers. While some fields required no sprays, others were sprayed just as many times as conventional cotton.[21] Yields, too, were variable and, with more sprays than they had budgeted for, some farmers complained they had paid a 'Rolls Royce' premium for something that ran more like 'a Holden'.[22]

A platform for integrated pest management?

While some rural industry groups hope transgenic Bt crops will smooth the way for a broader-based IPM in cotton, the reality may be somewhat different. 'IPM means different things to different people', the Australian Cotton Conference was told in 1996. For example, while organic farmers try to maximise biological controls to minimise the exposure of non-target organisms (such as humans and beneficial insects) to toxic chemicals, the cotton industry has—to a large extent—based its approach on rotating the use of broad spectrum insecticides.

While biological controls such as food sprays are being researched in Australia today for use on cotton, they require further work before they can be considered for widespread use. And by the time they are ready, it may be too late—at least for Bt. Food sprays patented by Rhône Poulenc have been hailed as the perfect biological control complement to Bt cotton, with the proviso that 'Bt resistance does not quickly muddy the water'.[23] Food spray researcher Robert Mensah of NSW Agriculture admits that it will be a challenge to complete research before resistance develops to Bt cotton.[24]

The food spray system involves the use of strips of lucerne throughout the cotton field to lure green mirids away from the cotton. The lucerne also attracts beneficial insects which can in turn be lured into the cotton crop using a food spray, when the time is right, to prey on heliothis. The food spray also repels heliothis; researchers claim that sprays can be reduced from seven to one with the same yield achieved.[25]

If Bt cotton is to be a platform for IPM, something will have to give. To embrace food sprays, for example, farmers would have to move away from monoculture farming and broad-spectrum sprays which disrupt the biological diversity required by the system. In short, it will take more than a transgenic crop to move farmers towards a broader-based IPM. Social factors will be just as important in shaping the future of this technology.

The social context of Bt transgenics

A 'pesticide driven mindset' (the 1996 Cotton Conference was told), encouraged by the relatively cheap cost of sprays, may be one of the biggest threats to the Australian cotton industry. It may also undermine those producers seeking to introduce a truly diverse range of pest management tools. Conventional cotton farming, some argue, needs a paradigm shift. 'Attitudes must change! The idea that "the only good bug is a dead bug" must give way to an appreciation of the ecological role of a pest species and acceptance of their presence.'[26]

A familiar sentiment has been echoed by the late Elaine Brough, a Queensland-based entomologist who advocated that producers overcome the risk of crises in pest management by moving away from a 'control focus' to a 'management perspective'.[27] Her concern was that 'the promotion of new biotechnological tools as environmentally safe and effective threatens to displace IPM programs and cause a return to reliance on a single pest management method'. Whilst suggesting that biotechnology had a lot to offer pest management, she argued that we might be 'stepping off the chemical pesticide treadmill onto the biotechnological treadmill'. Colleague Myron Zalucki also believes Bt cotton is seen as a 'silver bullet' to substitute failed chemical control agents. 'The current view of IPM is sample, spray and pray it works until the next tech fix comes along.'[28] This mindset has been encouraged by the language of Monsanto sales documents for Bt cotton which have offered farmers 'peace of mind knowing that crops are continuously protected'.[29]

Fundamentally, the biological component of IPM requires the careful understanding and manipulation of complex ecological interactions. It is an approach that so far has not tended to fit easily with the agribusiness imperative to develop patentable technological solutions. Rhône Poulenc's food spray product is one thing, the people skills involved in making the whole system work is another. Says Zalucki of IPM, 'It's as much people management as pest management', and he doubts we currently have the management systems in place to cope with such a sophisticated approach.

In the 1996–97 season, for example, Bt cotton itself required 50 per cent more time to check than conventional crops, and agricultural consultants could hardly meet the demand.[30] Some may well wonder if the industry is suitably equipped to cope with a labour-intensive biological control system based on food sprays, especially for transgenic Bt crops where large-scale regional co-ordination is essential.

The fact that Bt has now been engineered into around 40 other crops awaiting commercial release is an important factor here. The introduction of Bt genes into crops makes up 95 per cent of the commercial research into biological insecticides. Apart from cotton, Bt has been engineered into most of the major food and cash crops such as tobacco, corn, apple, canola, potato, rice, and tomato. While cotton is currently the only commercial transgenic Bt crop in Australia, Bt potato and corn have already been approved in the US.

Significantly, some crops such as corn and cotton share the same pests. This means that cornfields that might previously have provided populations of susceptible insects to dilute the resistant insects developing on cotton crops might, instead, provide more resistant insects! In other words, the growing numbers of Bt crops mean that we could wear out our Bt genes even quicker than forecasted. While some argue cotton is the most deserving of Bt genes,[31] incredibly there has been little dialogue between the different industries that use Bt, little indication of leadership from regulatory authorities, and even less public consultation.

The treadmill continues

Monsanto has said it wants to prevent resistance from 'diminishing the effective life span' of its product, and CSIRO scientists working with the cotton industry are hopeful that resistance management will extend Bt's 'useful shelf life'.[32] This very language, however, speaks of a disposable technology—with refugia-based resistance management seen as a tool to 'buy time', until the next chemical or transgenic fix comes along. One anxiously awaited development is a new variety of cotton which has two insecticidal genes 'stacked' into it—to act as a double hurdle to the development of resistance. One such 'pyramided variety' being developed by CSIRO contains two different varieties of Bt toxin genes. Researchers have warned, however, that even this could be of little use if resistance to the single-gene variety develops and knocks out one of the 'hurdles' in advance. As little as three years of careless use of the current variety could lead to this unfortunate situation.[33] Some researchers would have preferred Bt cotton not to have been commercially released until a second Bt gene had already been successfully engineered into the plant, but the cotton industry was desperate and 'just couldn't wait'.

Since 1998, the area of Bt cotton grown in Australia has roughly doubled to 165 000 hectares in 2000. Performance of Bt cotton has been variable and inconsistent, with one industry spokesperson describing it as 'cost-benefit neutral'. At the end of the 1998-99 growing season it was said that while many farmers were happy with its performance, equal numbers were 'extremely disappointed'. Right now, however, although it might be not performing as originally promoted, the industry says the transgenic crop has given it some breathing space with something other than chemical sprays to rely on. Most recent evidence shows that Bt cotton required 40 per cent less insect sprays than conventional crops, with other insects including mirids and aphids becoming more of a problem.

While researchers say there is no indication of resistance being a problem in the field so far, there is further evidence that undermines the assumptions upon which the refuge strategy has been built. CSIRO entomologists estimate that the insect resistance genes exist at a much higher levels in the field than previously assumed—'threateningly high', as one said. Also, Bt resistance appears not to be recessive which means it will take less exposure to the toxin for resistance to build up. Lastly, resistant insects have been found to emerge some two weeks after those from the refuge crops, severely limiting the chance of mating between the two to dilute out resistance.

At the 2000 Cotton Conference, there was much talk of the need for 'discretion and vigilance', praise for the 'caution taken in introducing the transgenic technology', and warnings that plantings should remain 'restricted' if the insect resistance genes were not to become a 'major threat'. The current Bt cotton crop now comprises 30-35 per cent of the total cotton acreage in Australia. The cotton industry officially supports recommendations from researchers that the area planted should be capped at this level until a two-gene variety of Bt cotton becomes available, although there are many with a 'shorter term vision' who would like to see the cap set at 70 per cent of total acreage.

Yet, there has also been a setback in delivery of the much sought after two-gene variety of Bt cotton, since the first attempt showed agronomic problems in US trials. According to Australian researchers these problems were not as serious under Australian conditions. However, the owner of the gene— Monsanto—did not allow the variety to be commercially released. This was much to the angst of the Australian cotton industry which says it would have been prepared to take the risk of using it. CSIRO is now working on another two-gene variety with a new Monsanto Bt gene, which it says will be due for commercial release in 2003.

There is also the promise of new classes of synthetic insecticides that can help 'buy time' until two-gene Bt cotton is available, although according to one commentator these insecticides do not have the potential for a sustainable environmentally friendly industry.[34] Whichever way it goes, the pressure seems to be on for the treadmill to continue.

Despite evidence that nature is proving more complex than anticipated, genetic engineers are confident their craft will solve all problems. Critics argue that biotechnology is flawed because of its attempts to reduce nature to small, definable pieces—such as genes—subject to human manipulation and control (see Will's critique of reductionism in chapter 3). They observe this has become the dominant scientific framework of explanation and inquiry to the exclusion of important other ways of knowing.[35] Applied to this case study, it appears that biotechnology, rather than sustainable practices as a whole, are 'signified' as the cutting edge of agricultural research. In adopting this argument, Jane Rissler of the US-based Union of Concerned Scientists says, 'If, in the last 40 years, we had spent as much in this country on sustainable agriculture research as we have on chemical research, we would have a lot of the answers.'[36]

Conclusion

With proponents claiming the age of chemicals is blending into the age of biotechnology, they predict a bright future for agriculture. Despite the sales hype, though, we have very little assurance that one of the 'flagships' of agro-biotechnology—Bt cotton—will provide a long-term sustainable future for the cotton industry.

On the one hand there are the cotton producers, caught between the rhetoric of integrated pest management and the reality of crisis pest management. On the other, the agricultural biotechnology industry, caught between the problems of resistance, and the need to recoup their investment. While Monsanto says it would be prepared to withdraw Bt from sale to prevent resistance, history has shown that chemical firms such as Monsanto have been more than willing to provide a procession of 'disposable' technological-fixes. Given the realities of the marketplace, and our relative ignorance of ecological processes, this could be Monsanto's preferred path.

Many farmers who have chosen transgenics may see little option but to continue on the treadmill of crisis pest management, but the impact of 'wearing out' Bt genes will go far beyond *their* farm gate. Farmers of edible food crops such as vegetables may end up using less benign products, and the increasingly profitable emerging organics industry will have lost one of the few insecticides they can use.

Who thus will benefit, and for how long, from Bt crops? While some may insist we are on the brink of another 'green revolution', will agriculture under the impact of genetic engineering really move to a more sustainable phase? If the last 'green revolution' is anything to go by, we may be at best simply leaving one agribusiness treadmill for another. Meanwhile, the appropriation of Bt by the transgenic juggernaut calls for every possible ounce of commitment to IPM by industry and government, with a strong co-ordinating hand by informed regulators to ensure a regional and cross-industry strategy for its use. This is a tall order, indeed, in the current era of free market economics.

11

Enclosing the biodiversity commons: bioprospecting or biopiracy?

JEAN CHRISTIE

Australia is the only developed megadiverse country which has a strong science and technology capability. However, despite this capability, Australia does not access or use its genetic resources to any great extent. Instead, Australia has tended to access and use exotic genetic resources ...[1]

The Western system is seeking to replace or destroy the communal regimes of intellectual property rights. It is the pharmaceuticals industry which stands to gain most from indigenous knowledge regarding biodiversity. From the earliest explorers to the present, Australia's native plants have been subject to investigation for their commercial potential.[2]

Australia occupies a unique place in the 'North–South' debate about biodiversity and who can use it. Though sitting squarely with the cash-poor biodiversity-rich countries of Africa, Asia, and Latin America that comprise the biodiverse 'South' geographically, Australia aspires to play in the big league of corporate biotechnology with the cash-rich, biodiversity-poor, industrialised 'North'.[3] In the quest for industrially useful biodiversity, Australia has cast its lot firmly with the plunderers and resource users of the industrialised North, in a growing political debate about who shall have access to and control over the earth's biological resources, and on what terms. Yet Australia is more likely to be plundered than be a plunderer; more likely to be resource provider than resource user.

Australia is already a target in industry's search for commercially useful biological resources. Corporate bioprospectors from Japan, Europe, and the US have set their sights on Australia's biological bounty. Gambling that its own biotechnology capacity will someday grow to compete with these transnational giants, Australia has aggressively adopted the stance of the industrialised North in international negotiations about 'access to biodiversity' and intellectual property over living organisms. Setting itself up as a junior partner to industrial giants, Australia has opened itself to corporate commercial exploitation, ignored the ethical questions associated with 'life patenting' in the biotechnology industry,[4] and offered scant protection to Aboriginal people and Torres Strait Islanders, whose knowledge of biodiversity will increasingly become the target of bioprospectors from within Australia and around the world.

Biotechnology is a multi-billion dollar industry, and is expanding daily.[5] Yet despite a growing armory of sophisticated techniques at their disposal, biotechnologists must still rely on the biological wealth of the earth for their 'raw materials'. Animals, plants, micro-organisms, and their microscopic parts are—and will remain—the foundation of the life industries. These 'genetic resources' may come from farmers' fields, the fishing and hunting grounds of indigenous peoples, state lands, national parks, deep-sea trenches, or thermal vents. They may be used essentially as they are found, their proteins isolated or purified, or their genes bioengineered from one species into another. They may be put to one of hundreds of possible industrial purposes. Whatever their source or eventual use, however, living organisms remain the essential life-blood of the new biotechnology industries. Without them, the industries themselves would die.

This simple fact is at the geopolitical heart of the world's interest in 'biodiversity'. Industry's need to control living organisms fuels a growing controversy over whether living things should be subject to 'intellectual property' claims. And it explains why biodiversity and intellectual property control over living organisms have become the subject of heated debate in two important international agreements signed in the 1990s: the Biodiversity Convention and GATT (the General Agreement on Tariffs and Trade), especially the latter's agreement on Trade Related Aspects of Intellectual Property Rights (TRIPS) now administered by the World Trade Organisation (WTO). Of further impact is the WTO's

Agricultural Agreement, and the Multilateral Agreement on Investment. Both erode national sovereignty and extend the rights of transnational industry.

Bioprospecting, biopiracy & intellectual property

There is no escaping the fact that the world's biodiversity, however it is measured, is overwhelmingly found in the tropics and sub-tropics of Africa, Asia, Latin America, and Australia—thousands of kilometres from the major life industries of North America, Europe, and Japan. The South therefore stands to gain, or lose, a great deal in the global race underway to control and profit from the earth's biological resources. The statistics on biodiversity speak volumes. Tropical forests contain at least half of all known plant and animal species worldwide. A fifty-hectare area in Malaysia may contain 830 native tree species, while all of Europe north of the Alps has only fifty. There may be more plant species in Botswana, Lesotho, Namibia, South Africa, and Swaziland than in any other region of comparable size in the world. Biological diversity in the deep ocean rivals that of tropical rainforests, and new research reveals that deep-sea species-diversity is richest in the tropics, diminishing as one moves toward either pole. Coral reefs are home to more species than any other marine ecosystem.

Over half of this diversity is in the western Pacific, and about 15 per cent in the tropical Atlantic. The Indo-Western Pacific has an estimated 1500 fish species, and over 6000 molluscs, compared with only 280 fish and 500 mollusc species in the eastern Atlantic. Tropical Lake Malawi has 245 species of fish; Lake Windermere in the UK has only nine. Brazil has over 3000 freshwater fish species—three times more than any other country in the world. Indonesia ranks first, second, sixth, and eighth in the number of endemic (or unique) species of birds, mammals, reptiles, and amphibians respectively, while Britain has few or no endemic species in any of those four categories.[6]

The potential economic value of all this biodiversity is staggering. Not surprisingly, agricultural biodiversity is also far richer in the South than in the temperate North. Every one of the world's main crop and livestock species has its centre of greatest genetic diversity somewhere in Africa, Asia, Latin America or the Middle East, in the areas where they have been longest cultivated. Apart from an emerging 'bush tucker' industry which is beginning to capitalise on the rich food heritage of Australia's indigenous peoples, virtually all of Australia's agricultural commercial production depends on 'exotic' (foreign) genetic stock. To a significant extent, this also is true for all other 'developed' countries of the North. Quite literally, the gene pool that world agriculture must rely on is kept alive and developed by the daily work of small family farmers of the 'developing' countries. Their crop varieties, adapted over centuries to thousands of unique ecological niches, hold the genetic secrets of disease and pest resistance, and adaptability to climate change, that industrial plant breeders and farmers must rely on for years to come. The Rural Advancement Foundation International (RAFI) has estimated that the annual value added to industrialised country agriculture just by farmers' seed varieties held in the gene banks of international agricultural research centres is US$4–5 billion.[7]

The commercial implications of 'traditional knowledge' and biodiversity are even more dramatic when one considers the world's pharmaceutical trade. For instance, 57 per cent of the 150 most frequently prescribed brand name drugs in the US in 1993 were derived from natural sources,[8] and the percentage is expected to rise with the advent of modern biotechnology. In 1996, global pharmaceutical industry sales were already a whopping US$222 billion.[9] Medicines from natural sources are a growing multi-billion dollar industry.

Western scientists have barely begun to assess the medicinal potential of the world's biological resources. Fewer than 15 per cent of Australia's species, for instance, have even been described by modern science.[10] But the world is rapidly waking up to the fact that most knowledge about the earth's known biodiversity is held in the minds and cultures of the indigenous and rural peoples who have used it for millennia. Nobody is learning this faster than industrial scientists. 'Traditional knowledge', once thought to be irrelevant to the industrialised world, is acquiring a new value—as industry grasps the importance of indigenous peoples' science, and ethnobiologists are hired to scour the globe in search of living organisms whose properties have been known and used for generations.

Once these properties are 'discovered' by modern science, and described in Western terms, industry increasingly uses intellectual property laws to appropriate both the biological resources and the knowledge about them as private property. Community innovators who may have created the 'resources', and led scientists to them in the first place, go unacknowledged and unrewarded. By legal sleight of hand, the living organisms themselves are deemed inventions of human intellect, and become the subject of monopoly patent claims. And here the fundamental conflict arises, and accusations of 'biopiracy' are made.

In most indigenous and rural societies, knowledge and innovation are not seen as commodities but as community creations, handed on from past to future generations. The earth and nature are used, developed, analysed, and managed, but are not exclusively owned. By contrast, European-based intellectual property regimes, and the international conventions that have taken their lead, are founded on the belief that innovative ideas and products of human genius should be legally protected as private property. In the last few years, intellectual property has evolved to encompass living organisms. Plant breeder's rights and recent applications of patent law increasingly cover a vast array of living things that are considered products of human genius, and subject to private monopoly controls.[11]

In the 1990s, and now at the dawn of the twenty-first century, these conflicts of worldview and values have come into sharp focus in the international negotiating arena. Indigenous peoples, farmers' organisations and environmentalists, with some vocal scientists, public sector researchers, public interest groups, ethicists, and academics from around the world have challenged industry, and called for 'No Patents on Life', arguing that life forms (including human genetic parts) should not be privatised by intellectual property claims. In the rarefied atmosphere of international conventions, these issues are being hotly debated, because intellectual property is increasingly seen as the newest mechanism for the industrialised countries of the North to gain monopoly control over the resources of the South, with the blessing of international law.

Two international agreements: the biodiversity convention & 'TRIPS'

The International Convention on Biological Diversity (or Biodiversity Convention) was adopted at the Rio de Janeiro 'Earth Summit' in June 1992. It affirms that nations have sovereignty over the biological resources within their borders, but requires that 'intellectual property' over living organisms must be respected. It also recognises the importance of 'traditional knowledge' in the conservation and use of biological diversity. Hidden in these concepts are a multitude of contradictions, biases, and inequities (see Table 11.1 below for details relevant to intellectual property and indigenous knowledge).

Six months after the Biodiversity Convention came into force (in December 1993), the 'Uruguay Round' of GATT was signed at

Table 11.1 The Biodiversity Convention (intellectual property and indigenous knowledge)

The Convention on Biological Diversity—a multilateral agreement—had been signed and ratified by 169 governments as of mid-1997. These governments (notably excluding the USA, which has not ratified the Convention) are the 'contracting parties' in the text below. The extracts are especially relevant to biodiversity, indigenous peoples' knowledge, and intellectual property rights. Because the Convention is an agreement among states, any protection offered to indigenous and rural communities is therefore via sovereign states.

Preamble, point 12: [Recognises] the close and traditional dependence of many indigenous and local communities embodying traditional lifestyles on biological resources, and the desirability of sharing equitably benefits arising from the use of traditional knowledge, innovations and practices relevant to the conservation of biological diversity and the sustainable use of its components.

Article 1, Objectives: The objectives of this convention ... are the conservation of biological diversity, the sustainable use of its components and the fair and equitable sharing of benefits arising out of the utilisation of genetic resources, including by appropriate access to genetic resources and by appropriate transfer of relevant technologies, taking into account all rights over those resources and to technologies.

Article 2, Use of Terms, point 13: 'In situ conservation' means the conservation of ecosystems and natural habitats and the maintenance and recovery of viable populations of species in their natural surroundings and, in the case of domesticated or cultivated species, in the surroundings where they have developed their distinctive properties.

Article 3, Principle: States have ... the sovereign right to exploit their own resources pursuant to their own environmental policies, and to ensure that activities within their jurisdiction or control do not cause damage to the environment of other States ...

Article 8, In-situ Conservation, clause (j): Each Contracting Party shall ... (j) subject to its national legislation, respect, preserve and maintain knowledge, innovations and practices of indigenous and local communities embodying traditional lifestyles relevant for the conservation and sustainable use of biological diversity and promote their wider application with the approval and involvement of the holders of such knowledge, innovations and practices and encourage the equitable sharing of the benefits arising from the utilisation of such knowledge, innovations and practices.

Article 10, Sustainable Use of Components of Biological Diversity, clause (c): Each Contracting Party shall ... (c) Protect and encourage customary use of biological resources in accordance with traditional cultural practices that are compatible with conservation and sustainable use requirements.

Article 15, Access to Genetic Resources, clauses 1, 4, 5, 6: (1) Recognising the sovereign right of states over their natural resources, the authority to determine access to genetic resources rests with national governments and is subject to national legislation ... (4) Access, where granted, shall be on mutually agreed terms ... (5) Access to genetic resources shall be subject to prior informed consent ... (6) Each Contracting Party shall endeavour to develop and carry out scientific research based on genetic resources provided by other Contracting Parties with the full participation of, and where possible in, such Contracting Parties.

Article 16, Access to and Transfer of Technology, clauses 1 and 2: (1) Each Contracting Party, recognising that technology includes biotechnology ... undertakes ... to provide and/or facilitate access for and transfer to other Contracting Parties of technologies that are relevant to the conservation and sustainable use of biological diversity and make use of genetic resources ... (2) ... In the case of technology subject to patents and other intellectual property rights, such access and transfer shall be provided on terms which recognise and are consistent with adequate and effective protection of intellectual property rights ...

Article 17, Exchange of Information, clauses 1 and 2: (1) The Contracting Parties shall facilitate the exchange of information ... (2) Such exchange of information shall include exchange of results of technical, scientific and socio-economic research, as well as information on ... indigenous and traditional knowledge as such and in combination with the technologies referred to in Article 16 ... It shall also include ... repatriation of information.

Article 18, Technical and Scientific Co-operation, clause 4: The Contracting Parties shall ... encourage and develop methods of cooperation for the development and use of technologies, including indigenous and traditional technologies ... [and] shall also promote co-operation in the training of personnel and exchange of experts.

Article 19, Handling of Biotechnology and Distribution of its Benefits, clause 2: Each Contracting Party shall ... promote and advance priority access on a fair and equitable basis by Contracting Parties ... to the results and benefits arising from biotechnologies based upon genetic resources provided by those Contracting Parties.

Marrakech, Morocco. On 1 January 1995 the World Trade Organisation came into being, to administer and monitor the GATT and the ongoing process of global trade 'harmonisation' which it called for. For the first time in history, a global trade accord contained explicit obligations for signatory states to adopt intellectual property laws, including monopolies over living organisms. This came to be known as the 'TRIPS Agreement'.

Taken together, the Biodiversity Convention and the WTO's TRIPS Agreement demand enactment of laws and practices that have profound ethical and economic implications for all governments and peoples. Article 27 of TRIPS, for instance, obliges all signatory states to introduce intellectual property laws over 'life'. Specifically, TRIPS requires WTO members to provide for patents on micro-organisms, and to have some form of intellectual property over plants. 'Animals' may be excluded from patentability, though it should be noted that in practice the microscopic parts of animals including humans are treated as micro-organisms. Governments' failure to introduce such intellectual property laws could be construed by the WTO as an unfair trade barrier, and the 'offending' country would then be subject to trade sanctions.

Article 16 of the Biodiversity Convention stipulates that Western-style intellectual property regimes must be respected. As governments try to come to grips with the implications of these two agreements, indigenous peoples and other non-government groups and organisations worldwide are questioning both the ethics and the benefits of their intellectual property requirements.

While the Biodiversity Convention and WTO's TRIPS have the force of international law, governments however have considerable flexibility in interpreting and implementing the intellectual property clauses. Furthermore, the time frame for their implementation is either undefined, or gives leeway to developing country legislators well into this first decade of the twenty-first century. When TRIPS was adopted, it was agreed that its controversial Article 27.3(b) would be reviewed in 1999— one year before developing countries were to have implemented the provision. But the deadline for implementation in the South has come and gone, and a substantive review has not yet been done. The WTO is now arguing that any review that does take place should be limited to questions of implementation, not substance. Yet many WTO members, including most African countries, have made proposals for substantive changes. Though it will take a monumental effort, the agreement could still be changed in ways that strengthen the hand of the South.[12] Indeed, policy makers, non-government and indigenous peoples' organisations, farmers, and environmentalists from all over the world could well influence the direction of national and global policies on intellectual property over living things, and peoples' knowledge about them.

Public policy considerations: four case studies

The following examples illustrate some of the thorny issues that must be addressed by international bodies and policy makers in all countries (like Australia) which have ratified and are now bound by the terms of the Biodiversity Convention and the World Trade Organisation's TRIPS Agreement. They highlight a range of public policy considerations that are being debated around the world, and indicate some of the competing interests that must be taken into account.

Case study 1: Western Australia smokebush patented in the United States

Between the 1960s and 1981, the Western Australia Herbarium assisted a licensed collector for the US National Cancer Institute (NCI) to collect over 1200 plant specimens from WA, for a program to screen natural products for their cancer-fighting properties. All were sent to the NCI for testing as anti-cancer drugs. Among the specimens were samples of a WA smokebush (*genus Conospermum*). In the late 1980s, the specimens were screened again by the NCI on behalf of the US National Institutes of Health (NIH).[13] A compound named conocurvone was extracted and purified from the smokebush and analysed for its effect on HIV. Even in low concentrations, it was found to 'inactivate' the virus.[14] Anticipating a significant commercial use, the US—represented by the Department of Health and Human Services—applied for a US patent in January 1993 (granted September 1997, patent number 5672607) involving 24 claims on antiviral naphthoquinone compounds, compositions, and uses thereof. Conocurvone was one of these compounds.

The following year, the same applicant applied for an Australian patent (granted August 1997, patent number 680872). Like all patents it gave its US owner exclusive monopoly rights to use the compounds from the WA plant, to decide who would be licensed to use them, and at what cost. This case was widely reported in the Australian media, and fuelled a public debate that mirrored growing worldwide concern about who should own, have access to, and benefit from the world's biological resources.

The patent application prompted hastily conceived and controversial amendments to the *WA Conservation and Land Management Act*, which now gives the State environment minister power to grant exclusive rights to the State's flora and forest species for research purposes.[15] It also led to agreements in December 1993 and March 1994 between the US National Cancer Institute, the Western Australian State government, and the Australian Medical and Research and Development Corporation (AMRAD)—a Victoria-based pharmaceutical development company.[16] AMRAD was granted an exclusive worldwide licence by the NIH to develop and market a series of anti-HIV drugs derived from the conocurvone,[17] and AMRAD paid the Western Australian government an initial A$1.15 million to ensure access to other smokebush species.

It is unclear what role, if any, the Australian federal government played in these negotiations. Nor has anyone publicly acknowledged that Aboriginal people have used the smokebush medicinally for centuries.[18] Nobody has revealed whether its prior use by Aboriginal people led the NCI to the smokebush in the first place; one might well ask. One might also ask whether Aboriginal and Torres Strait Islands peoples will have any protection in future cases where their medicinal plants and knowledge lead to drug discoveries, or whether State governments should have the right to sign away the biological heritage of the Australian continent for private profit.

Case study 2: AMRAD—a 'made-in' Australia bioprospector & broker

AMRAD was established in 1986 for 'the discovery, development and commercialisation of pharmaceutical programs and projects'. Its 'members' include eleven medical research institutes from five Australian States and Territories. In 1993, AMRAD established a natural products screening program, and has since 'secured access to biota from some of the most diverse environments in the world—from the wet tropics to Antarctica'.[19] Within Australia, AMRAD has signed collection agreements with government authorities, research institutes, State herbaria, botanical gardens, Aboriginal Lands Councils, and Land Trusts.

The company is collaborating, for example, with the Darwin-based Menzies School of Health Research and the Tiwi people of Bathurst and Melville Islands, on research into 'twelve plants still used as bush med-

icines'.[20] Agreements like this one, and another in 1995 for the pharmaceutical screening of Arnhem Land Plants signed with the
(Aboriginal) Northern Land Council, have sent reverberations through
Aboriginal Australia which is still struggling to weigh up the implications of such precedent-setting arrangements, and to assess how best to
protect indigenous peoples' rights in the face of corporate approaches.

In the early 1990s, AMRAD signed deals with foreign-based
transnational companies, including Merck Sharpe and Dohme (USA),
Kanegafuchi Chemical Industry Company (Japan), and Sandoz (now
Novartis, of Switzerland).[21] In 1995, it also agreed to screen Australian
natural products for the Chugai Pharmaceutical company of Japan[22] and
in May 1997 announced a five-year, A\$15 million, agreement with
Rhône-Poulenc Rorer (France) to screen Australian plants for use in the
treatment of asthma and related illnesses.[23] In 1997, AMRAD also
signed an agreement with the state government of Sarawak in Malaysia
for access to Sarawak's genetic resources.[24] By April 1997, AMRAD
had tested over 750 000 biological extracts for use in treating such conditions and illnesses as HIV, cancer, hepatitis B, neurodegeneration,
anaemia, and asthma.

Who, one might ask, will reap the benefits from these agreements—
indigenous peoples, other Australians, or corporate coffers? Will the
customers of any new medicines be able to buy them at reasonable cost,
or will they pay inflated monopoly prices? On what terms were the
agreements signed? Will indigenous peoples be adequately recognised
or rewarded for their contributions? Will their knowledge and intellectual integrity be protected?

Case study 3: Andean grain patented in the US & Australia

For centuries, the indigenous peoples of the Andes have grown, eaten,
and adapted quinoa (*Chenopodium quinoa*) to hundreds of micro-envi-
ronments in their arid mountain homelands. Quinoa—one of the most
protein-rich grains in the world—has been a staple of Andean diets for
generations. It went almost unnoticed in the rest of the world until
recently, when the food industry started to take an interest in it. The
Swiss food manufacturing giant Nestlé has been conducting research on
quinoa, and has been granted a quinoa-processing patent in the USA.[25]
Quinoa has begun to appear in 'exotic' breakfast cereals and pastas, one
of a number of 'ancient grains' now attracting the attention of health-
conscious consumers in the industrialised world.

This might offer a chance for Quechua and Aymara quinoa producers of Bolivia, Chile, Ecuador, and Peru to develop a sustainable export into the markets of the industrialised countries to their north—but not if current trends continue. Two scientists from Colorado State University in the US have been granted US and Australian patents on male sterile plants of a Bolivian quinoa variety named Apelawa, and any quinoa hybrids produced with them.[26] These patents give the inventors the right to prevent anyone else from making, using or selling quinoa hybrids derived from the patented Apelawa cytoplasm, without permission or payment of royalties. Legally the inventors also have the right to prevent hybrid quinoa from entering the US, if it has been created using their patented technology. The same scientists, incidentally, are conducting hybridisation research on Peruvian quinoa varieties, which, according to one of them, sparked commercial interest.[27]

Bolivia's National Association of Quinoa Producers has asked the patent holders to drop the Apelawa patents, and has protested about them at home and at the United Nations. Andean governments are contemplating action. If these patents are not dropped, and if new quinoa patents are granted, the intellectual property rights for commercial development of quinoa in the North will likely wind up in the hands of corporations, and the patent holders will reap the benefits of the patent monopolies over this age-old Andean food crop. The economic implications of this precedent are obvious for farmers of all crops the world over.

Case study 4: Indigenous knowledge for the taking?

In the beautiful grounds of the Dreamtime Centre in Rockhampton, Queensland, Aboriginal, and Torres Strait Islander guides take visitors daily on walking tours of the Centre's expanding collection of native plants, whose leaves, bark, roots, fruit or flowers have been used since time immemorial by Australia's first peoples—for food, fibre, medicinal, and other purposes. Each plant is identified by its scientific name, and by its use. Each reflects the ingenuity and skill of the people who for centuries have studied, described, and used the many species of trees and shrubs, vines and creepers which are now displayed in the living exhibit of the Centre's garden.

The Dreamtime Centre's botanical collection is part of an important and growing movement to document, study, and preserve indigenous peoples' knowledge of nature. At the moment, however, there is nothing but goodwill and an ethereal promise of recognition and the 'repatriation of information' under the Biodiversity Convention to stop the world's bioprospectors from simply appropriating these resources and the knowledge that goes with them. Since early colonial times in Australia, European missionaries, and botanists have compiled records of the medicinal uses that Aboriginal peoples have made of native plants.

In more recent times, a burgeoning bibliography of books and studies has been compiled to record the unique and often endangered knowledge of peoples who are losing their lands, their cultures, and their historical relationships with biodiversity. Gene hunters need only pay the price of admission to gardens like the Dreamtime Centre's, borrow the books from public libraries, conduct research on the plants so identified, and then patent their results. They need not give so much as a nod to the people whose knowledge led them to the plants in the first place, and shortened their research time immeasurably. Such a scenario is not just alarmist speculation, it is absolutely consistent with present-day trends.

The life industries, for instance, are already approaching botanical gardens all over the world, seeking access to their plant collections.[28] Deals are already being struck—with the assistance of a loophole in the Biodiversity Convention which says that all biological collections made before the Convention came into force are outside its scope. Corporations can therefore side-step the spirit of the Convention, which obliges them to negotiate with the source countries of biodiversity, by going instead to cash-strapped botanical gardens whose officials may be more willing to make a deal.

Bioprospecting or biopiracy?

These four case studies are among thousands of examples from around the world which illustrate what has come to be known as 'bioprospecting'—the search for commercially useful biological resources. Private companies, academics, and public sector researchers (increasingly in partnership with a corporate backer) are combing the fields, forests, waters, and traditional medicine chests of the world, for useful biological resources. In reality, 'biopiracy' more accurately describes what they have done to date. Resources and knowledge have simply been appropriated from countries and peoples whose permission was seldom sought, and who received little or no credit or compensation for their contribution to bioindustrial developments. A move is now afoot to correct this practice.

The case studies presented above also illustrate some of the pressing public-policy concerns that all interested parties must address, including:

- how to assess the impact of national sovereignty over biological diversity on different interests—including various levels of government, indigenous peoples, industry, and public sector institutions;
- how to regulate access to biological diversity, and determine the conditions under which access is granted; defining what is meant by an 'equitable sharing of benefits' and finding ways to ensure that it happens, both nationally and internationally;
- assessing the implications for different interests of intellectual property monopolies over living organisms; and
- conserving the knowledge of indigenous peoples' and other communities; protecting the intellectual integrity of indigenous peoples in the face of intellectual property regimes.[29]

Conclusion

Australia prides itself on being the 'clever country', and biotechnologists are seen to be among the cleverest of all, leading Australian industry into the new millennium. But just how clever are they? And clever at what cost? The answers will lie in how Australia tackles the policy issues above. By consistently arguing the case for industrial interests, will Australia develop a viable biotechnology industry of its own, or set itself up as a perpetual branch plant to Northern corporations—providing raw materials but rarely owning and controlling production? In aspiring to the big league, will Australia simply duck the ethical questions about whether living organisms (including human genetic material) should be patentable, or will it limit the scope of patent law? Will Australia subvert the collective rights of Aboriginal and Torres Strait peoples under its present intellectual property regime, or use its obligations under the Biodiversity Convention to contribute to the process of reconciliation? In the very near future we are certain to discover the answers, especially with heightened public awareness and debate about the issues of bioprospecting and biopiracy.

12

Opposing genetic manipulation: the GeneEthics campaign

BOB PHELPS

At the dawn of the twenty-first century, many key events have occurred which have the potential to end the short and chequered history of genetically modified organisms (GMOs)—or what some of us prefer to call genetically *mutilated* organisms.

During 1999-2000, the environmental and public health impacts of the industrial and therapeutic uses of gene technology were repeatedly exposed, leading to an expansion in public opposition to GM crops and foods globally. It became clear that, in overseas nations, industry and government had misinformed the public over food-related disasters like L-tryptophan and mad cow disease (BSE). Government failure world-wide to require the labelling of GM foods heightened suspicion about, and rejection of, these new foods. The blanket use of public relations disinformation posing as education—by industry and governments to fast-track GMOs onto farms and our dinner tables—backfired, and intensified public mistrust and retailer scepticism. The more people have learned about GMOs the less they seem to like them.

Farmer and rural industry confidence and support for genetic engineering also began to wane as corporate domination of the global food and fibre supply intensified, as companies repeatedly made unauthorised releases of GM crops at secret locations, as GM crops gave lower yields and were less profitable than the companies had promised, as genetic pollution of conventional and organic crops by GM food and fibre products occurred, and as insurers said any damage done by GM crop use may be uninsurable. Many investors took money out of agri-cultural gene technology, forcing most top GM companies to merge, be taken over, or to make plans to sell their agbiotech operations.

In contrast, support for GM-free futures grew strongly through increased public demand for healthy, natural, and safe organic food; through the expansion of sustainable, environmentally friendly, syn-thetic-chemical-free agriculture; and through increased demand and premium prices for GM-free and organic foods.

Origins of the gene technology industry

The so-called life science firms—Aventis, Monsanto, Dupont, Novartis to name some in the top tier—are the same agrochemical companies (but renamed) which brought destructive poisons into farming fifty years ago. Chemical use is now widely discredited, but aggressive pro-motion and weak regulation still allows many new, poorly tested, and loosely monitored poisons to enter the environment and food supply.

Genetically modified organisms on farms would further intensify destructive chemical-based farming systems. The so-called Gene Revolution is a form of industrial bio-roulette which will further erode the diversity of food crops and reduce food security world-wide. Designer organisms are developed mainly in the interests of corporate capital, and are being promoted with minimal public participation, con-sultation, or acceptance. Their future is advanced by misinformed, myopic, and over-enthusiastic governments as the preferred future for agricultural production.

Gene-spliced organisms entrench chemical/industrial agricultural practices and extend corporate monopolies by modifying organisms for intensive factory and farming systems. One example is that of GM foods designed to have a longer shelf-life to allow their transport to more prof-itable and distant markets. Products in the mass consumer markets must arrive at the point-of-sale looking presentable and fresh, even if their nutritional and health value may be compromised. Their conventional counterparts would have rotted long before. Such high-tech foods cannot feed the world's malnourished, as claimed by the bioindustry, because the poor cannot afford foods which attract premium prices—to cover the extra royalty, and transport and handling costs. The genetically engineered long-life FlavrSavr tomato was the first attempt to design a vegetable that never appeared to age. It was sold in US shops in 1994.[1] Yet, many of these tomatoes were soft and bruised on arrival, despite special packaging, and so could not be sold as fresh fruit.[2] Calgene, which owned the FlavrSavr, collapsed and was sold to Monsanto. 'Counterfeit fresh' tomatoes seem to have now been abandoned.

In nature and in traditional farming systems, almost all new varieties are created within species' boundaries. Nature rigorously tests their capacity to survive, so many organisms fail and disappear. The survivors are the plants, animals, and microbes that successfully adapt and are productive, so only environmental successes are reproduced. Gene technologists simply attempt to tear up nature's rulebook, in an effort to ensure that the most profitable, but not necessarily the fittest, survive.

Many genetic engineers see nature's limits as mere hurdles to be jumped or avoided. They cut and paste genetic codes between organisms that could never mate in nature, turning all living entities into mere factory fodder. Organisms are reduced to isolated genes that are scrambled in test tubes, disregarding the complex systems in which all life exists. Through genetic manipulation, 'foreign' DNA is repeatedly and randomly fired imprecisely into unknown locations in target organisms, to hopefully create modified organisms of commercial interest. For example, a bacterial gene for herbicide tolerance has been transferred into potatoes and soybeans so that the crop can be sprayed with weed killer more often, at higher doses and less carefully, to kill weeds better. The environment and buyers of polluted food are the losers.

Corporate & investor confidence

The GM industry has come to rely on false promises rather than performance to sell its products. These lies are now being exposed, and subsequently the industry is being strongly opposed and appears in decline. Genetic engineering will reduce, not increase our capacity to feed and clothe the world, protect and restore the environment, and nurture public health. The enormous resources being poured into genetic engineering could be utilised better if redirected into the development of integrated, biodiverse, organic management systems.

The central promise of modern biotechnology to feed the world is like many other public relations scams, created merely to assuage public concern and encourage acceptance of dangerous and inappropriate technologies. The benefits have been strongly promoted, while the costs, hazards, and risks have been downplayed or ignored, despite abundant warnings of harmful impacts well before these technologies have been deployed. The promised pay-offs of mega-technologies have never mate-

rialised or have been short-lived, and many impacts are still not officially acknowledged.[3] This generation and future ones have already been
saddled with radioactive wastes and toxic chemicals, some of which
will degrade health and the environment for millennia. A similar pattern
of reckless use is being encouraged by lavishly funded government and
corporate promotional campaigns extolling the (false) promises of a
bioengineered world.

In its 1999 budget, the Australian government allocated $10 million
to Biotechnology Australia (BA), located in the Department of Industry
and Science. Of this, $4.4 million was for its so-called public awareness
program, to ameliorate public concern and promote acceptance of GM
(see also chapter 2). A further $3 million was injected into this program
in the 2000 budget, as part of an allocation of $30.5 million over four
years for the government's National Biotechnology Strategy. A total of
about $15 million per year for three years, including State government
and private contributions, was for assisting companies to 'bridge the
commercialisation gap'; $3.65 million was 'for assessing the require-
ments and costs involved in segregating products which have been
developed with gene technology'; and the smallest amount of all—$0.25
million—was 'for developing an environmental risk research program.'[4]

Biotechnology Australia retained the services of the transnational
public relations firm Turnbull Porter Novelli to assist with its program
of regional forums where critics of GM are generally excluded from
official participation. During 2000, 2.2 million of BA's promotional
leaflets on GM foods were distributed through supermarkets nationally,
and two sets of materials have been distributed to all high schools
nationally. Biotechnology Australia has also funded many meetings,
publications, and displays by industry proponents, while denying financial support to critics.

Despite all this window dressing, in the final week of November
2000 the NASDAQ index of high technology stocks fell to its lowest
level for a year. Aventis, a Franco-German drug and agrochemical company (formed earlier in 2000 by a merger of AgrEvo and Rhône
Poulenc) announced its intention to sell its agricultural business by the
end of 2001. This followed huge unauthorised GM releases of Aventis
crops and food in North America and Australia. The company's seed
business shrank by Euros 3.5 billion during the year.

Another big factor in the general downturn for GM was a European
backlash against GM foods and weak demand everywhere else.
Agrochemical and seed giants, AstraZeneca and Novartis, merged their

agribusinesses into the GM conglomerate Syngenta. This was a stock-market flop, when investors valued the float at just over half of the expected US$10 billion. Swiss drug company Pharmacia bought out Monsanto to acquire its drug, artificial sweetener, and related assets, and is expected to sell the agricultural remnant within two years. Smaller GM players DuPont and BASF were expected to sell their GM drug-making businesses but keep their agricultural assets.[5]

Insurance companies are in the business of accurately assessing and hedging against risk to make a profit. So another serious blow to the industry occurred when the Insurance Council of Australia and one of the big three insurers worldwide—Swiss Reinsurance Rueck—warned that GM would be difficult to insure and that insurance companies would be loath to accept the risk. This reluctance is due to uncertain and difficult-to-define risks; potentially very large impacts and compensa-tion; and because the bioindustry has alienated the general public. The Insurance Council of Australia said insurers would be cautious as 'GM is dangerous (and) characterised by an extremely diversified risk pro-file of a new technology,' and 'there have been many pharmaceutical disasters where parallels may be drawn to GM technology'.[6] Significantly, during 1999, Europe's largest bank, Deutsche Bank, issued two reports advising institutional investors to steer clear of agri-cultural GMOs.

The GeneEthics Network campaign

The Australian Conservation Foundation (ACF) began its genetic engi-neering campaign in January 1988, with part of the royalties from the Midnight Oil band's album *Species Deceases*. The campaign structure changed in 1991, when the Australian GeneEthics Network was formed as a separate entity, under ACF sponsorship, to build a coalition of groups and individuals to work for strong laws on genetic engineering; critical public debate and education; and public control to prevent envi-ronmental, social, health, economic, and ethical impacts issuing from genetic engineering.

GeneEthics plays a leading role in promoting education, solutions, and actions to enable informed public participation in deciding which food and farming futures we will embrace. In 1988, the Network 'uncovered' unauthorised releases of transgenic pork into Adelaide's food supply, and publicised the attempted cover-up in the national news media (see also chapters 2 and 9).[7] In 1992, we held public awareness and discussion forums about the issues of genetic engineering in all capital cities and many regional centres. *The Troubled Helix*, published in 1992 and again in 1994, is a widely used set of case studies and resources for schools that examine the environmental, social, and ethical issues of gene technology. Our website was established in 1996. *Habitat* (Australia) education supplements published in 1999 and 2000 have aided the tide of public opposition to GM.

From 1992, the Network received small annual federal government grants for its education programs until the Liberal/National coalition was elected to government in March 1996. Support ended abruptly, but the generosity of thousands of loyal Network supporters has enabled the campaign to continue. Activities have included mobilising large numbers of activists to demand GM food labelling and GMO regulation, and to oppose GMO releases. The emailing of newsletters to thousands of subscribers has been another activity. The Network has also been engaged in policy analysis and development, lobbying policy-makers and government departments, mounting public awareness and education campaigns, liaising with the news media, publishing education and campaign materials, organising and participating in conferences and public debates, and staging protests.

Since 1998 the Network has campaigned strongly for a minimum five-year freeze on GMO releases and for GM-free zones. A freeze on all gene technology and its products is needed while government establishes several important safeguards including:

- a stronger Office of Gene Technology Regulator (OGTR) with enforceable national licensing laws;
- comprehensive and mandatory labelling of all foods produced using gene technology;
- GM-free zones where organic and conventional produce can be grown and local councils can establish GM-free food services in crèches, and 'meals on wheels';
- an effective biosafety protocol to protect biodiversity and public health when GMOs are transferred internationally (now agreed to and expected in 2002);
- an end to patents on life which enable corporate domination of the global food supply; and
- an enforceable liability and insurance regime to ensure compensation for any GM damage, and to encourage corporate responsibility.

Around Australia and in many countries overseas, such as New

Zealand (which now has a one-year freeze on all GMO releases and a Royal Commission of Inquiry), South Africa, the US, and UK, State and local governments have declared GM-free zones. At the State level in Australia, this includes Tasmania (a one-year freeze on all GMO releases and a public inquiry) and Western Australia (which adopted a two-year freeze on commercial releases, and the State Labor Party took a five-year-freeze policy to the 2001 election). Many local councils such as Waverly, Moreland, Willoughby, North Sydney, and Byron have GM-free food services, while four councils in WA, at least one in Victoria, and several in NSW have declared themselves opposed to GMO releases.

The Network mounts joint campaigns with other groups. These include the Public Health Association of Australia, which researches and lobbies on food safety; the Friends of the Earth anti-genetic engineering collective, which focuses upon corporate, environmental, and social justice questions; *EcoConsumer,* which researches and publicises food-industry programs and policies, and regulation by the Australia New Zealand Food Authority (ANZFA); and the Australian Consumers' Association (ACA), which convened a Consensus Conference on GM food in 1999 and campaigns for GM food labelling. Public opinion polls reflect the Network's and the ACA's stand on the labelling of GM foods. A poll conducted in July 2000 for the the *Sydney Morning Herald* by the national Nielsen organisation indicated that 93 per cent of the Australian population supported labels; 65 per cent would avoid GM food if possible; and 58 per cent did not want GM drugs.

In addition, the Network has extensive links with groups overseas which campaign against GMOs. These include Greenpeace International (Holland), No Patents on Life (Germany), the Pure Food Campaign (US), the Third World Network (Malaysia), the Pesticide Action Network Asia-Pacific (Malaysia), the Research Foundation for Science, Technology and Ecology (India), the Genetics Forum (England), and the Rural Advancement Foundation International (Canada). The highly influential Global Days of Action, held in thirty countries during April and October 1997, opposed GM foods, factory farming, and life patents.

GM regulation

The Gene Technology Act 2000, which established the Office of Gene Technology Regulator to assess, license, and monitor GMOs, was passed on Friday, 8 December 2000 in the final moments of the Commonwealth parliamentary session. The law was rammed through over the many objections and positive amendments proposed by Greens, Democrats, and Independents in the Senate. The Liberal/National government and the Australian Labor Party (ALP) voted as one in the Senate, where the government did not have a majority. The GM industry, federal government, and opposition thus colluded to enable GMOs and their products to be fast-tracked into the environment and food.

Since 1988, the Australian GeneEthics Network has sought mandatory, independent, and precautionary notification, assessment, licensing, and monitoring of all GMOs—especially those proposed for release into the open environment. The goals of the Network have not yet been achieved, as few of the numerous formal and informal submissions delivered to government and its inquiries have been implemented. The Network sought to amend the government's Gene Technology Bill 2000 in the Senate. One major concern was that the Bill could not ensure that all proposals to release GMOs were, without exception, notified, assessed, licensed, monitored, and insured. It gave broad exemptions to a range of uses of GMOs including human genetic engineering, and animals for organ and drug production. For example, 600 transgenic goats with human albumin genes being created by the South Australian government's Research and Development Institute (SARDI), under contract to a Chinese drug company, need no licence and can be approved by SARDI's Institutional Biosafety Committee. All uses of GMOs, gene technology, and their products should be prohibited unless assessed and licensed.

Major changes to the Gene Technology Bill required the co-operation of the ALP opposition, which alternatively could have voted with the smaller parties and independents to form a majority in the Senate. The Network wrote to and lobbied shadow cabinet ministers, asking them to use the ALP's Senate numbers to enact as a minimum the recommendations of the lead-in Senate Inquiry Report, *A Cautionary Tale: Fish Don't Lay Tomatoes*, published in November 2000. The report was prepared by the Senate Community Affairs Committee, chaired by ALP Senator Rosemary Crowley. While the report's recommendations were weaker than was desired, at least they resulted from a public consultation process, and included proposals that had emerged before in the 1992 report of the earlier Australian government's inquiry into genetic engineering (see chapter 2 for a history of regulation in Australia). GeneEthics argued that:

- environmental impact assessments on GMOs should be required under the trigger provisions of the Environment Protection and Biodiversity Conservation Act 2000 (EPBC Act);
- the OGTR should be a statutory authority rather than a single regulator;
- critics as well as applicants be accorded rights of appeal against OGTR decisions;
- liability and compensation be required where GMOs contaminate conventional or organic crops or products; that license conditions require insurance;
- the GMO and GM product register include details of the license holder and all their agents, as well as the exact location, status (commercial/experimental), and scale (hectares/litres etc) of all uses of GMOs;
- each OGTR committee—scientific, community, and ethics—have equal resourcing and agendas; that States be enabled to provide a higher level of environmental and public health protection than the minimum set by the Act; and
- States and local government be enabled to ban all or some uses of GMOs for a limited time or permanently.

The ALP worked with government, the States, and industry to ensure passage of the Bill, ignoring most of the suggestions from GeneEthics. The Bill conformed to the government's and the opposition's globalisation, trade, and science agendas. One partial success however was an agreement reached outside the Act that will allow States to remain selectively GM-free for up to four years, when the Act is to be reviewed.

A watered-down version of the precautionary principle in the object of the Act also partly fulfils the ACF's 1989 policy goal that, 'The onus should be on the proponents of genetic engineering applications to reveal all the potential hazards of their proposals and to show that the uses are benign, not on governments or citizens to demonstrate the dangers.'[8] In the Gene Technology Act 2000 the precautionary principle is restricted by a requirement that it must also be cost effective. It is also unclear whether the principle will apply to the licensing provisions of the law, the mechanism for ensuring company compliance. Though the Gene Technology Act 2000 was due for review in 2005, GeneEthics mobilised Network supporters to seek amendments to the system during 2001.

At the international level of GM regulation, the Network has lobbied Australian diplomats over the last five years for a tough Biosafety Protocol. The protocol negotiated under the Convention on Biological Diversity is to regulate transnational movements of GMOs (for example, for purposes of importation, trade, scientific research, etc). The text of the Cartagena Protocol (named for the city in Colombia where the penultimate negotiation was held) was agreed to on 29 January 2000 in Montreal. It was followed by a meeting of the Intergovernmental Committee for the Cartagena Protocol (ICCP) in France from 11-15 December 2000, to begin implementing its provisions on a web-based Biosafety Clearinghouse Mechanism, where information about transfers, approvals, and so forth will be posted, and where capacity building will be provided for developing countries.

The protocol will come into force once fifty countries have signed and ratified it. The Australian government is under pressure from many industry groups not to sign until it becomes clearer how the protocol will be implemented. The general expectation is that the protocol may come into force in 2002.

GM food standards & labelling

The Australia New Zealand Food Standards Council (ANZFSC), composed of the ten health ministers of Australia and New Zealand, agreed in August 1999 that Standard A18 would require all foods, food additives, and processing aids produced using gene technology to be labelled. But US government and industry pressure on prime minister John Howard and State premiers led to the watering-down of the health ministers' final decision, adopted on 28 July 2000.

The standard will require the labelling of food and food ingredients only where novel DNA and/or novel protein is present in the final food, or where ANZFA decides the food has altered characteristics. The standard will come into force one year after publication of its final text in the government gazette, which means there will be no GM labels before 2001 (if ever). Exempt from GM labelling are:

• highly refined food (such as sugars and oils), where the effect of the refining process is to remove novel genetic material and/or novel protein;

• processing aids and food additives (yeasts, enzymes, setting agents, thickeners), except where novel genetic material and/or novel protein is present in the final food;

- flavours which are present in a concentration less than or equal to 0.1 per cent in the final food;
- food prepared at point of sale (such as restaurants, hotels, and 'take-away' food locations); and
- foods where any one ingredient contains up to 1 per cent of GM food, where a paper trail that accompanies the ingredients shows it presence to be inadvertent.[9]

This means that many of the GM foods now being approved by ANZFA will be regarded as 'substantially equivalent' to conventional foods, and will go unlabelled. This includes nineteen different kinds of imported soy, corn, canola, sugar beet, potatoes, and local and imported cottonseed products, foods from plants engineered with such things as Bt toxins (insect killer), those with viral particles (virus resistant), others with antibiotic resistance marker genes (in most engineered plants as part of the production process), and those derived from plants with herbicide tolerance (leaving up to 200 times the present allowable level of chemical residues). Consumers might be concerned!

Small amounts of many of these products may have been in processed foods since December 1996, when the first shipment of GM soy reached Brisbane. Since 6 May 1999, when Standard A18 on GM foods was added to the Food Standard, they have also been *provisionally* approved pending final assessment by ANZFA.

There is a zero threshold of contamination for any food which makes a claim to be GM-free, including organic foods, which need not make a specific label claim as the Organic Standard excludes GM foods. According to the Australian Consumer and Competition Commission (ACCC) it is likely under provisions of the Trade Practices Act 1974 concerning false and misleading claims, that anyone who labels 'GM-free' will be the target of testing. Any detectable contamination with a GM ingredient, or any use of GM in the food's production, may lead to prosecution. This means that the GM food industry is putting the integrity of Australia's GM-free food supply in doubt, then unfairly transferring the costs of quality assurance onto GM-free products.

The food industry will be required to document and/or test the GM status of their ingredients, and to label accordingly. Genetically modified ingredients will appear in the ingredients panel of a label, with the words 'genetically modified'. Single-ingredient GM foods, such as a packet of GM soy beans, will require the words 'genetically modified' to appear against the name of the food on the front of the packet. This does not adequately protect public health, safety, or the consumer's right to know. Foods produced using gene technology have no history of safe use but ANZFA does not undertake pre-market human testing or post-market monitoring. General release of novel GM food products into the food supply makes all Australians and New Zealanders part of the giant uncontrolled r-DNA experiment. The potential for new genes and proteins to create allergic reactions or other illnesses is poorly understood and poses a threat, according to the Public Health Association and many other informed commentators worldwide.

Foods are assessed and approved on a case-by-case basis but assessments are based on the unscientific assumption that most GM foods are 'substantially equivalent' to ordinary foods. This focuses on the superficial similarities of end products rather than the basic, potentially dangerous differences between conventional and r-DNA production processes and their impacts. Substantial equivalence was dreamed up by industry to support the hoax that GMOs are not radically different from traditional breeding where the transfer of genetic material between unrelated or distantly related species is rare.

Controlling 'runaway' companies?

The agrochemical giants have poured tens of billions of dollars into GMO development, and want a return on investment. But few people appear to be enthusiastic purchasers of their products. While farming groups believe that a biotech future will assist Australian agriculture to be more competitive, the commercial survival of food GMOs is in serious doubt. It is desperation time, and in the haste to get GMOs into the environment and onto our plates the companies have failed time and again to comply with laws and guidelines. The practical impossibility of segregating GM from GM-free products, and controlling GM pollen and seed in the environment, is now clear.

Canadian seed producer Advanta was caught out early in the year when canola seed exported to Europe and grown in four countries was found to be about one per cent GM, though certified as GM-free. The plants were removed and destroyed. It appears that the accepted separation distances for the production of uncontaminated seed are inadequate in Canada where GM varieties are extensively grown. Advanta now intends to move its seed production operation to Eastern Europe, Australia or another location that may remain GM-free.

In February, farmer and GeneEthics Network activist Leila Huebner found herbicide-tolerant GM canola plants—from an Aventis CropScience site near Mt Gambier in South Australia—uncovered in a roadside dumpster and on the local tip. This flouted GMAC's minimal guidelines. Only after a journalist followed the story did the interim OGTR act on Ms Huebner's complaint. Aventis had not even informed the farmer growing the offending canola that the crop was genetically engineered, referring to the experimental plants as 'hybrids'. Local council and neighbouring farmers were inadequately informed of the plantings. The OGTR report found there was lack of compliance at many Aventis trial sites in four States and that most plantings, on more than 1000 hectares, were not scientific trials as the seed was exported to Canada for commercial use. This was the fourteenth unauthorised release of GMOs in Australia since GMAC was formed in 1988, according to their annual reports, but all of those responsible had been excused from any penalty (see also chapter 2).

US giant Monsanto was discovered to have ginned about 60 tonnes of GM cotton, tolerant to the herbicide Roundup, produced in field trials. It was mixed with conventional cotton and sold for food, animal feed, and fibre in contravention of GMAC guidelines. Roundup Ready cotton was approved later in 2000 for general commercial release, and Monsanto expected around 15 000 hectares to be planted in the 2000/2001 season. In November 2000, Monsanto reported to the OGTR that it had sold enough Bt cotton seed to plant 20 000 hectares over the approved quota of 165 000 hectares. The quota is 30 per cent of the entire cotton crop, set by the regulators as part of an insect resistance management strategy which tries to ensure that insects do not acquire Bt resistance in response to the huge amount of Bt toxin that GM plants pour continuously into the soil and open environment (see also Salleh's account in this volume). The National Registration Authority on Agricultural and Veterinary Chemicals (NRA), which registered the Bt cotton plants as a pesticide, has given Monsanto a dispensation to lift the limit to around 180 000 hectares this season. Some seed was recalled and some of the crop ploughed in to meet the revised quota, but this major breach of GMAC guidelines has otherwise also gone unpunished. Monsanto's over-supply of Bt crops poses a hazard to the environment-friendly and sustainable use of Bt microbes as a natural insect spray, which poses little insect-resistance risk. This unauthorised release threatens the livelihoods of other users of microbial Bt (especially organic growers).

On 18 September 2000 in the US a coalition of public interest groups released lab results which showed the presence in Taco Bell taco shells of the StarLink variety of GM corn, not approved for human consumption. StarLink produces a *Bacillus thuringiensis* (Bt) insecticidal toxin called Cry9C, which, because it is unable to be digested, had been assessed as a potential human food allergen. After public outrage, Aventis agreed to buy back the entire year's corn harvest and to sell it for animal feed and ethanol production. Many farmers didn't even know they had grown StarLink but 350 flour mills around the country had received shipments. It is doubtful that StarLink was kept out of the human food supply, showing how difficult it is to contain genes once they are out. Segregation of GM and GM-free crops and their products in the food chain is an unachievable goal when a large variety of planters, combines, augers, grain elevators, trucks, mills, storage bins, silos, and other facilities are all in use for both product streams. Massive recalls, plant closures, and government buy-backs as the result of the StarLink fiasco have for the first time shaken US public confidence in the safety of GM foods, which are all currently unlabelled.

If commercial release of GMOs proceeds in Australia, then emerging evidence of secondary GM impacts upon bees, honey quality, monarch butterflies, ladybugs, lacewings, and various other flora and fauna should be conclusively resolved. CSIRO Entomology has begun a long overdue, three-year, A$3 million research project into the risks of commercial GM crops. GeneEthics has argued for a freeze on all GM releases while the study is completed, using the abundant data now available for analysis from extensive commercial experience with GM crops and foods in North America.

The Gene Technology Act 2000 enables the OGTR to determine whether applicants for a licence are fit to hold it or not. The environmental and compliance records of Monsanto, Aventis, Novartis, DuPont and other GM companies suggest they should not be trusted with licenses for GM crops under the new law.

Sustainable agricultural systems

The GeneEthics Network assists groups actively engaged in sustainable farming or which advocate the replacement of chemical/industrial agriculture with sustainable systems. Such groups include the Organic Federation of Australia, NASAA, ORGAA, the Biodynamic Research Institute, Permaculture, Biological Farmers of Australia, the Diggers' Club, Eden Seeds, the Heritage Seed Curator's Association, the Seed

Savers Network, as well as allergy and sensitivity groups, feminist groups, doctors, natural therapists, animal rights groups, concerned scientists, and many others.

The Network asserts that, were it adequately researched and resourced, the traditional organic farming systems used by farmers can improve farm productivity, maintain crop diversity, and provide major benefits to the world's population. In other words, if appropriately supported, modern organic agriculture can deliver nutritious foods while guaranteeing food abundance and security.

Certified organic farming systems which use no synthetic chemicals have rejected GMOs since the early 1990s, and also prohibit food irradiation. Organic foods are therefore a safe choice. To maintain diversity and achieve sustainability in agriculture, the ACF GeneEthics Network argues that 'adequate government funding should be provided to encourage research into alternatives to genetic engineering technologies.'[11] At present, research and development (R&D) into sustainable agriculture in Australia receives less than one per cent of government funds for agricultural R&D. Over $80 million per annum is going into agricultural GMO R&D while the organic industry gets less than $0.5 million per annum.

As much of public sector science is now required to have business support, funds increasingly go to technologies that are tied to patents and profits, not to producing integrated environmental, social, and economic systems. Benign management systems such as organics, herbal medicines, and open-pollinated crop varieties are marginalised and unfunded because they cannot be easily monopolised by big business interests.

Sustainable systems such as organic farming and agroecology that work in harmony with nature are expanding rapidly despite minimal R&D funding, few tax breaks, and low official acceptance (see also chapter 9). By supporting organic agriculture, food buyers pay the true cost of growing 'clean and green' food, keeping more Australians on the land, and nurturing our environment. Organic agriculture creates biologically active soil, teeming with microflora and organic matter, to produce the micronutrients that healthy plants require. Organic farmers use the latest research data in systems design and soil science to keep plants and their environment in balance, so crops are more resistant to pests and diseases.

Genetically modified organisms address the symptoms not the causes of past rounds of environmental damage. The degradation created by current rural management practices and technologies will be compounded by GM technology. For example, salt tolerant plants cannot fix salination, drought tolerant crops would enable more marginal lands to be exploited, and larger and faster growing GM fish have the capacity to destroy natural fish stocks. Many of the problems in rural ecosystems could be fixed by organic agriculture that takes a complete system approach. Community support for the processes and products of sustainable farming, and opposition to the introduction of GMOs can ensure future generations a secure food and fibre supply.

Conclusion

The ACF GeneEthics Network and kindred groups want to ensure that Australians have the option of a GMO-free future. Demand worldwide is strongest for GM-*free* food and fibre products. Premium prices are being paid for them. Once lost, our GM-*free* status will be gone forever as GMOs and their genes cannot be recalled to the laboratory. Recent events involving GMOs clearly indicate they have no future in food and farming.

Despite this, industry and governments are determined to force the proposed Bio-utopia upon us, whether or not we like it. Community control is overdue on the allocation of public money for education, research, and development, and promotion of all our future options so that genuine choice is restored. All foods produced using GMOs must be labelled and a 'One Stop Shop'—a unified, precautionary system of regulation— should be enacted.

A five-year freeze on GMO releases, at the very least, is urgently needed to stop the fast tracking of GMOs, to assess the success or failure of GMOs in North America and the few other countries using them, to fully inform the community about all our food options, and to assess the costs of GMOs to the environment and to public health. A freeze would allow us to consider the best ways to implement sustainability in Australian rural industries, including ways to foster a creative partnership between rural and urban people on this most important matter.

Activists have worked tirelessly over many years to ensure Australia's food and farming industries will have a bright future. Much remains to be done. What must be created are new options that challenge an *altered-genes* future, and which serve the real needs of Australians for ecologically sound foods, fibres, materials, and sustainable production systems for this and all future generations.

Endnotes

Introduction

[1] Griffiths *et al.* 1996

Chapter One: Bio-utopia

[1] Turner 1988
[2] Kaku 1998, p. 5
[3] Kaku 1998, p. 9
[4] McAuliffe & McAuliffe 1981, p. 8
[5] Suzuki & Knudston 1988, p. 116
[6] see Cribb 1994, p. 9
[7] see Cribb 1994, p. 9
[8] Ho 2000
[9] Haynes 1994, p. 97
[10] Bud 1993, p. 54.
[11] Huxley 1977, p. 22
[12] Huxley 1977, p. 14
[13] *Sydney Morning Herald* 20 October 1997, p. 11
[14] see Morton 1997 p. 798
[15] see the video, 'Le Clonage: Un Saut Dans L'Inconnu' ('Cloning: A Leap into the Unknown'), screened on ' The Cutting Edge', SBS TV 9 March 1999
[16] *Sydney Morning Herald* 20 October 1997, p. 11
[17] Wheale & McNally 1988, p. xvi
[18] see Rifkin 1984
[19] see George 1976; Hobbelink 1991; Shiva 1997, pp. 211–12
[20] O'Brien 2000
[21] Alibek 2000, p. 261
[22] Brook 2000, p. 1
[23] see BioTech Sage Report (http://www.biotechnav.com/)
[24] Australian Biotechnology Association 1999
[25] James & Krattiger 1996
[26] IOGTR 2000a, p. 13

[27] Woodford 2000

[28] The Brisbane Institute 2000, p. 40

[29] see Hindmarsh 1999

[30] *Sunday Mail* 2000

[31] Barboza 2000

[32] The People's Caravan was co-ordinated by the Pesticide Action Network-Asia Pacific (PANAP) Penang, Malaysia

[33] Venter & Cohen 1997, p. 28

[34] Jackson 2000, p. 1

[35] With regard to the Queensland Draft Code of Ethics on Biotechnology, see Hindmarsh & Hulsman 2000.

[36] Organic Federation of Australia 2000, p. 1

Chapter Two: Constructing Bio-utopia

[1] O'Neill 1990a

[2] O'Neill 1990b

[3] O'Neill 1990a

[4] Millis 1986, pp. 1–2

[5] Mussared 1994

[6] Mussared 1994

[7] see Hindmarsh *et al.* 1995; Hindmarsh 1996

[8] Hoad 1997, p. 40

[9] see Shapin 1988, p. 534

[10] Schattschneider 1960, p. 71

[11] Herman & Chomsky 1994, p. xi

[12] see Carey 1995

[13] see Horin 1976, p. 17

[14] see Krimsky 1982

[15] *The Australian* 1976

[16] Hansard 1977, p. 1

[17] Evans 1977, pp. 1-2

[18] see Wright 1986

[19] see Wright 1986, p. 598

[20] Hoad 1977

[21] Farrands 1979

[22] Brumfield 1980, p. 9

[23] Bartels 1984, p. 183

[24] Dredge 1990

[25] ACF 1990

[26] see Hindmarsh 1994, p. 382
[27] Lee 1992, p. 1
[28] Hindmarsh & Hulsman 1992
[29] Reeves 1990, p. 1; see also Peacock 1990; Stocker 1990
[30] see Hindmarsh 1992; Love 1993
[31] OECD 1992, p. 2
[32] see http://www.biotechnology.gov.au
[33] Darby & Metherell 2000, p. 1
[34] see ABC 2000
[35] Hindmarsh 2000

Chapter Three: Disrupting Evolution

[1] Judson 1979
[2] for example, Kauffman 1993; 1996; Webster & Goodwin 1996
[3] Such as the 'game of life' (Wuensche & Lesser 1992) or the sandpile model of 'self-organised criticality' (Bak 1996)
[4] cf. Dawkins 1986 & Lewin 1994
[5] Crick 1958
[6] see Crick 1970
[7] Lewin 1994, p. 3
[8] Dawkins 1986, p. 73
[9] see Lewin 1994 for a more complete explanation
[10] Nurse 1997
[11] Wills 1989
[12] Waldrop 1992; Kauffman 1993; Bak 1996
[13] Kauffman 1993
[14] Lövel 1997
[15] Wills 1996
[16] Prusiner 1997
[17] Kidwell & Lisch 1997; Shapiro 1997
[18] Wills 1996
[19] Wills 1996
[20] Wills 1994

Chapter Four: 'Get Out of My Lab, Lois!'

[1] Gilchrist 1995
[2] *SMH* (Sydney Morning Herald) 16 December 1995
[3] Smith 1995

[4] Nelkin 1987

[5] Kawar 1989

[6] Patterson & Wilkins 1991, p. 26

[7] Kawar 1989, p. 734

[8] Hornig Priest 1995

[9] Brown 1995

[10] Dunwoody 1993, p. 10

[11] Tiffen 1989

[12] Nelkin 1987

[13] Goodell 1986

[14] for example, see Mazur 1981

[15] see Loge 1991

[16] Ward 1995, p. 162

[17] Friedman 1986

[18] Goodell 1986

[19] Dunwoody 1993, p. 8

[20] Henningham 1995

[21] Henningham 1995, p. 94

[22] Dunwoody 1993, p. 20

[23] Dunwoody 1993, p. 42

[24] Dunwoody 1993, p. 28

[25] Ward 1995, p. 50

[26] Loge 1991, pp. 6–7

[27] Dunwoody 1993

[28] Dunwoody 1993, p. 41

[29] Loge 1991, p. 6

[30] Ward 1995, p. 164

[31] Hornig Priest 1995, p. 53

Chapter Five: The Quality-Control of Human Life

[1] Caplan 2000, p. 1

[2] see the video, 'Le Clonage: Un Saut Dans L'Inconnu' ('Cloning: A Leap into the Unknown'), screened on ' The Cutting Edge', SBS TV 9 March 1999

[3] Joyce 1990

[4] Rowland 1992; Schmidt 1997

[5] see O'Neill 1991

[6] see 'Le Clonage'

[7] see 'Le Clonage'

[8] Rowland 1992

[9] see 'Le Clonage'

[10] The Law Reform Commission of Victoria 1988, p. 15

[11] Dow 1997

[12] see Dow 1997

[13] see Saltus 1986

[14] see *Nature* 1992

[15] Wheale & McNally 1988, p. 257

[16] Elmer-Dewitt 1994, p. 24

[17] Schmeck 1986

[18] see Williams 1989

[19] The Law Reform Commission of Victoria 1988, p. 14

[20] McLaren 1987

[21] Davis 1984

[22] Saxton 1988, p. 218

[23] Saxton 1988, p. 224

[24] Beck 1995, p. 95

[25] see Hughes 1986

[26] O'Neill 1991, p. 43

[27] Katz Rothman 1986, p. 238

[28] see Vines 1986

[29] Wilkie 1996, pp. 133–4

[30] Hubbard 1988, p. 228

[31] Butler 1997

[32] Daniels & Murnane 1980, p. 129

[33] Lifton 1987, p. 16

[34] see Stanworth 1987, p. 30

[35] Stanworth 1987, p. 31

[36] Hubbard 1988, p. 232

[37] Wheale & McNally 1988, p. 274

Chapter Six: Genetic Testing

[1] cited in Hudson *et al.* 1995, p. 391

[2] Eder 1994, pp. 38–42

[3] see also O'Connor 1996, pp. 12–13

[4] Lloyd 1997, p. 26

[5] see Verhoef *et al.* 1996, pp. 5–16

[6] Guthrie Test: Originally, the test was used to screen newborn babies for phenylketonuria, a genetic disease, transmitted by simple

Mendelian recessive inheritance. The gene alteration leads to a toxic accumulation of phenylalanine and of its metabolites. The test uses a small droplet of blood obtained by heel prick spotted onto blotting paper and then analysed in a laboratory. Today, the blood spots are also used for screening for disorders such as cystic fibrosis, galactosaemia, and hypothyroidism.

[7] McEwen & Reilly 1994, p. 197

[8] Quade 1993, p. 14

[9] cited in Rothenberg 1997, p. 1756

[10] O'Connor 1996, pp. 70–71

[11] Pokorski 1995, pp. 13–14

[12] Masood 1996, p. 389

[13] Masood 1996, p. 391

[14] O'Connor 1996, p. 65

[15] Draper 1991

[16] cited in Rothenberg et al. 1997, pp. 1756–7

[17] O'Connor 1996, p. 86

Chapter Seven: Knowing Your Genes

[1] *The Australian*, 15 October 1997

[2] Gans 1997

[3] Connor 1997

[4] Myers 1990

[5] Hazen & Trefil 1990

[6] Wynne 1991

[7] Turney 1993

[8] Lederberg 1991

[9] Holub 1993

[10] *Time*, 25 April 1994, p. 48

[11] Marx 1993

[12] see http://www.genetests.org

[13] Maddox 1993

[14] *The Australian*, 19 March 1994

[15] *Time*, 1994, p. 26

[16] Carey 1989, p. 20

[17] Hazen & Trefil 1990

[18] Love 1993

[19] Wynne 1991

[20] Nelkin & Lindee 1995

[21] ABC-TV *Club* Buggery, 16 August 1997

[22] Prebble 2000

[23] Kingsnorth 1998, pp. 267–8

[24] *Weekend Australian*, 15–16 May 1999, p. 24

[25] see Beck 1995

[26] Dessaix 1994

[27] Holub 1992, p. 20

[28] Holub 1990a

[29] Jackson 1984, p. 122

Chapter Eight: Risks, Regulations & Rhetorics

[1] see Beck 1992

[2] Giddens 1991, pp. 122–4

[3] Neale 1997

[4] Hoy 1996a

[5] Uhlig 1996

[6] Brown 1997

[7] The Australian Food Council is now called the Australian Food and Grocery Council (AFGC) and is the peak national representative organisation for the Australian food, drink, and grocery products industry. It has a membership of more than 170 companies including subsidiaries and associates, representing in the order of 80 to 85 per cent of the gross dollar value of the industry. The AFGC's charter is to promote a domestic business environment conducive to international competitiveness, strong and sustained investment, innovation, business growth and profitability. This industry is Australia's largest manufacturing sector, and has an annual turnover in excess of $50 billion (see http://www.afgc.org.au/).

[8] see Hoy 1996a

[9] see http://www.afgc.or.au/documents/mr015_00.htm

[10] see Douglas 1992

[11] see Douglas 1966

[12] see Hannigan 1995, p. 169

[13] Charlesworth *et al.* 1989, p. 42. See also Gottweis 1995

[14] Zechauser & Viscusi 1996, p. 14

[15] Beck 1992, p. 70. See also Ho 1998

[16] see Crook 1999

[17] see IOGTR 2000b

[18] see Cabinet Office 1999

[19] Australian Biotechnology Association 1996
[20] Irwin 1997
[21] Wilkie 1996
[22] see Priest 1995
[23] see, for example, Charlesworth et al. 1989, p. 41; Gottweis 1995; Levidov 1996, p. 62
[24] The Australian Biotechnology Association 1996
[25] Barbano 1997
[26] European Federation of Biotechnology 1994
[27] Hoy 1996b
[28] Hoy 1996c
[29] www.biotechnology.gov.au
[30] see IOGTR 2000c, 2000d

Chapter Nine: Gene Technology, Agri-Food Industries & Consumers

[1] *Australian Farm Journal,* December 1998
[2] GMAC 2000
[3] see *Sydney Morning Herald,* 24 July 2000, p. 4
[4] quoted in *Australian Farm Journal,* December 1998, p. 38
[5] Lee 1992a; *Australian Farm Journal,* December 1998; Biotechnology Australia 1999
[6] Healy 1991
[7] Lawrence 1987; Lawrence 1995; Vanclay & Lawrence 1995; Gray & Lawrence 2001
[8] Sleigh 1988; Bunyard 1996; Rifkin 1998
[9] Kloppenburg 1991, pp. 488–9
[10] Biotechnology Australia 1999
[11] *Sydney Morning Herald* 24 July 2000, p. 4
[12] Burch et al. 1996; Lawrence et al. 1996; *Australian Farm Journal,* December 1998; Gray & Lawrence, 2001
[13] Organic Federation of Australia media releases 4 and 13 December 2000, www.ofa.org.au
[14] http://www.nfu.ca/frame.htm
[15] NFF 1993, p. 84
[16] Biotechnology Australia 1999
[17] *Australasian Biotechnology* 1991; Australian Food Council 1996; *Australian Farm Journal,* December 1998
[18] Bureau of Rural Resources 1991, p. 61

[19] Lawrence 1987; Kloppenburg 1988; Geisler & Lyson 1993; but also see discussion in Krimsky & Wrubel 1996

[20] *The Guardian* 1999

[21] Hindmarsh 1992; Steinbrecher 1996

[22] Yearley 1996, p. 42

[23] Lawrence et al. 1992; Gray & Lawrence, 2001

[24] *Ground Cover* 2000a

[25] see also *Sydney Morning Herald*, 24 July 2000, p. 4

[26] Lawrence & Norton 1994; Foster & Ghonim 1995; Lyons & Lawrence 1999; Norton 1999

[27] Lyons, Lockie & Lawrence 2000

[28] Burch, Lyons & Lawrence 1998

[29] Queensland Department of Primary Industries, media release, 'Organic Food Focus for Queensland', 4 July 2000

[30] Avery 1995

[31] Youngberg *et al.* 1993, p. 298

[32] Campbell 2000

[33] Allen 1993; Harper 1993; Nottingham 1998

[34] INRA 2000, p. 50.

[35] Wagner et al. 1997; Norton 1999; Kamaldeen & Powell 2000

[36] Crotty 1995; Duff 1994; Beardsworth & Keil 1997; Duff 1999

[37] INRA 2000, Norton 1999

[38] Norton 1999

[39] The draft standard requires labelling of food and food ingredients containing novel DNA and/or protein or where the food has altered characteristics. Exemptions include highly refined food (where the effect of the refining process is to remove novel DNA and/or protein); processing aids and food additives except those where novel DNA and/or protein is present in the final food; flavours which are present in a concentration less than or equal to 0.1 per cent in the final food; and food prepared at point of sale. The Standard also allows an ingredient to contain up to 1 per cent of unintended presence of genetically modified product (Australian New Zealand Food Standards Council Press Release 28 July 2000)

[40] Quotes from Australian Food Council 1996, p. 22

[41] *Ground Cover* 2000b

[42] Poulter 1999, ENN Direct 2000

Chapter Ten: Wearing Out Our Genes?

[1] Peacock 1995

[2] Fitt et al. 1994

[3] The bulk of research for this chapter took place between 1995 and 1997, and was accurate at the time of writing in 1997.

[4] NRA 1996a

[5] Avcare 1995

[6] McLean et al. 1997, p. 240

[7] CC 1996, p. 189

[8] Monsanto 1995

[9] McLean et al. 1997, p. 262; Verkerk & Wright 1997

[10] Forrester 1994

[11] Verkerk 1997

[12] NRA 1996b

[13] Fitt et al. 1994; Monsanto 1995; McLean et al. 1997, p. 259

[14] Forrester 1997

[15] In the US, 40 per cent of growers who planted the first commercial crop of Bt cotton had to spray to kill heliothis, contrary to what they had expected (Monsanto n-d). While the dose of toxin in Bt cotton was sufficient to control two US pests, it was insufficient to provide such effective control for another—*H. zea* (Prakash 1997; MacLean et al. 1997, pp. 259–72) (which, like the Australian heliothis pests, was naturally more tolerant to Bt). Some farmers suffered losses due to infestation from *H. zea* (Hutton 1997; MacLean et al. 1997, pp. 259–72). Fears are now held for the longevity of Bt's effectiveness against *H. zea* (Roush 1997a; Prakash 1997)

[16] see CC 1996, p. 73; Forrester 1997

[17] see CRDC 1997; Pyke 1997; McLean et al. 1997, p. 240

[18] CC 1996, p. 59

[19] Monsanto 1995

[20] Zalucki 1997

[21] CRDC 1997; Forrester & Constable 1997

[22] Landline 1997

[23] CC 1996, p. 76

[24] Mensah 1996

[25] CC 1996, pp. 147–51; Mensah 1996

[26] CC 1996, p. 153

[27] Brough et al. 1995

[28] Zalucki 1997

[29] Monsanto 1995

[30] CRDC 1997; *Australian Cotton Outlook* 1997, p. 1

[31] McLean et al. 1997, pp. 280–1

[32] see Monsanto 1995; CC 1996, p. 62

[33] McLean et al. 1997, p. 270

[34] Roush 1997a

[35] see Busch et al. 1991

[36] Quantum 1996

Chapter Eleven: Enclosing the Biodiversity Commons

[1] ARC 1994

[2] Henriette Fourmile, Burkal Consultancy, Cairns 1995

[3] Rural Advancement Foundation International (RAFI) (see website—http://www.rafi.ca) monitored negotiations of the United Nations Food and Agriculture Organisation and the Convention on Biological Diversity during 1995–97. Australia characteristically aligned itself with the US, and consistently adopted a pro-industry position

[4] see, for example, The CornerHouse 1997

[5] RAFI 1996a

[6] see Shand 1997, pp. 8–9

[7] RAFI 1994

[8] Grifo et al. 1997

[9] see Internet source: www.ims-international.com/press/corp96.htm

[10] Glanzig 1996

[11] see Hindmarsh 1999

[12] see Third World Network 2000. For a full review of TRIPs 27.3(b): An update on where developing countries stand with the push to patent life at WTO; March 2000, see http://www.grain.org

[13] US Patent Office 1997

[14] Armstrong & Hooper 1994

[15] Fourmile 1996

[16] Armstrong & Hooper 1994; Fourmile 1996

[17] AMRAD 1996

[18] Fourmile 1997

[19] AMRAD 1997a

[20] Sweet 1997

[21] see Hindmarsh 1994, p. 252

[22] Australasian Biotechnology 1996

[23] AMRAD 1997b

[24] AMRAD 1997a

[25] On 3 November 1993 Nestec (Switzerland) was granted US patent 5264231 for a quinoa processing procedure, and have filed for patents in Europe on the same process. Nestlé has expressed interest in quinoa research being conducted in Ecuador

[26] Duane Johnson and Sarah Ward hold US patent no. 5304718, and Australian patent no. AU 9222922.

[27] Ward 1997

[28] RAFI 1996b

[29] Marrie, A. pers. comm., 10 September 1997. Indigenous organisations have proposed that Australia develop *sui generis* legislation, consistent with indigenous expectations, to protect indigenous heritage, including indigenous peoples' knowledge of biodiversity. They endorse the concept of 'heritage' elucidated by Erica-Irene Daes, Special Rapporteur of the Sub-Commission on the Prevention of Discrimination and Protection of Minorities, in *Principles and Guidelines for the Protection of the Heritage of Indigenous Peoples* (1996)

Chapter Twelve: Opposing Genetic Manipulation

[1] Krimsky & Wrubel 1996, p. 100

[2] Gene Exchange 1997, p. 12

[3] see Hynes 1989; Beck 1995

[4] Undated letter from Senator Nick Minchin, Minister for Industry, Science and Resources

[5] *The Economist*, 18 November 2000, p. 125

[6] ICA (Insurance Council of Australia) submission to the House of Representatives Inquiry into Primary Producer Access to Gene Technology, 2000

[7] ACF 1990

[8] ACF 1989

[9] Australian New Zealand Food Standards Council Press Release 28 July 2000

[10] ACF 1989

Glossary

Entries have been drawn mainly from five main sources. We would like to thank the publishers of those sources for giving us permission to use the entries cited, including: Oxford University Press (*A Concise Dictionary of Biology*, 1990, 2nd edn; Stoddart Publishing Co. Ltd, Canada (Suzuki, D. and Knudston, P., 1988, *Genethics*; Zed Books (Shiva, V. and Moser, I., 1995, *Biopolitics*); Pan Macmillan Australia (Haynes, R., 1991, *High Tech: High Co$t*; Pluto Press (Wheale, P. and McNally, R., 1990, *The Bio Revolution: Cornucopia or Pandora's Box*). Where applicable, the key for entries is listed at the end of the glossary.

agrochemical Synthetic chemical used in industrialised agriculture for plant production (fertiliser) or protection (pesticide) (1).

allergy A condition in which the body produces an abnormal immune response to certain antigens (called *allergens*), which include dust, pollen, certain foods and drugs, or fur. In allergic individuals these substances, which in a normal person would be destroyed by antibodies, stimulate the release of histamine, leading to inflammation and other characteristic symptoms of the allergy (for example, asthma or hay fever) (+5).

amino acids The base chemical subunit or building blocks of proteins; there are twenty common amino acids (+3).

anthrax A malignant infectious disease of cattle, sheep, and other animals and of humans, caused by *Bacillus anthracis.*

bacillus A rod-shaped bacterium.

Bhopal The world's worst chemical accident. On 2 December 1984 noxious fumes containing methyl-isocyanate, hydrogen cyanide, and cyanide leaked out from the Union Carbide plant in Bhopal, India. Up to 16 000 people are thought to have been killed as a result of breathing the toxic fumes, another 125 000 suffered some health damage—at least 30 000 of whom are described as incurably ill. Union Carbide settled for US$470 million in 1991 and quietly left the country (see Rowell 1996, pp. 96-97).

biochemistry The study of the chemistry of living organisms, particularly the structure and function of their chemical components (principally proteins, carbohydrates, lipids, and nucleic acids) (+5).

biodevelopment Bioindustrial development.

biodiversity The millions of life forms found on earth, their genetic variety and variability, the ecological roles they perform, and the interrelated and interdependent ecological communities or ecosystems they form, or are found in (+1).

bioindustrial Industries engaged in the life sciences.

biology The study of living organisms, which includes their structure (gross and microscopial), functioning, origin and evolution, classification, interrelationships, and distribution (5).

biosciences The biological or life sciences include biochemistry, biomedicine, cellular biology, chemical engineering, microbiology, molecular biology.

bioscientists Scientists engaged in research and development in the life sciences.

biotechnology Development of products by exploiting biological processes or substances. Production may be carried out by using intact original or modified organisms, such as yeasts and bacteria, or by using active cell components, such as enzymes from organisms (4); the attempt to engineer and control biological processes for the purposes of industrial production (1).

Bio-utopia Is short for biological utopia; the Enlightenment scientist and philosopher Francis Bacon (1561-1626) is credited with advancing the 'scientific' notion of a biological utopia in his book *New Atlantis* (see, for example, Weinberger 1989, pp. 73-75; Krimsky 1991, pp. 84-85). His writings serve to justify the creation of new species (or new 'strains' of species) as liberating an oppressed humanity from hunger, disease, etc.—just as genetic engineers now, some 300 years later, advocate the purpose of genetic engineering. The morality of creating new species is primarily anthropocentric, or human-centred, that is Nature exists to 'serve' humans. In the extreme form of this morality, largely driven by an economic agenda, the environmental ethics of gene-altering are ignored, denied, or pushed aside.

blood plasma The liquid part of the blood (that is, excluding blood cells. It consists of water containing a large number of dissolved substances, including proteins, salts, food materials (glucose, amino acids, fats), hormones, vitamins, and excretory materials (+5).

Down's syndrome A congenital form of mental retardation due to a chromosome defect (+5).

bovine somatotropin (bST) A commercially manufactured

form—using genetically engineered microbes—of bovine growth hormone produced in the pituitary glands of cows essential to milk production. Regular injections of biosynthetic BGH—known as bST (or r-bST)—can raise the productivity of dairy herds. Use of r-bST is presently prohibited in Australia and the European Community, but is common in the USA. Anti-bST campaigners claim that its use is cruel because of the frequency of injections and increased risks of mastitis in cows, that it poses a risk to human health both as a possible carcinogen and through over-use of antibiotics (to treat mastitis), and that it represents an economic threat to small-scale dairy farms.

BRCA1 Breast cancer gene 1. Located on chromosome 17. The link between early-onset breast cancer to BRCA1 gene (for breast cancer gene 1) was first discovered by Hall et al. (1990). Soon the discovery of two other mutations, one called p53 (Malkin et al. 1990), the other BRCA2 (Wooster et al. 1994), on chromosome 13, followed. Women who inherit the mutated form of the gene have a substantially increased lifetime risk of developing breast or ovarian cancer.

carrier In some genetic mutations, such as cystic fibrosis, which affects about one in 2500 Australians, the genetic defect is unnoticeable in a carrier. In carriers only one of the two inherited copies of the gene is defective. The defect is masked by the second healthy copy, and such individuals do not suffer from the condition themselves. However, they can pass the defective copy on to their offspring. Thus, if two carriers produce children, each child has a one in four chance of being afflicted with the disease.

cell The structural and functional unit of all living organisms (+5).

cell culture Growth of cells under laboratory conditions.

central dogma: The belief of molecular biology that hereditary information generally flows unidirectionally from DNA molecules to RNA molecules to proteins (see also genetic determinism) (for a good explanation, see: Vandermeer, 1996, pp. 50-55; Ho, 2000, pp. 95-106).

CFTR Cystic fibrosis transmembrane conductance regulator gene; the genetic defect in cystic fibrosis lies in this gene. It has a role in the transport of ions; more than 400 mutations in the gene have been reported to be responsible for the dysfunctions, thus accounting for the striking variation in lung function between afflicted individuals, although other genetic and nongenetic factors, as yet not identified, may also influence the manifestation and prognosis of the disease.

chromosome The condensed rod made up of a linear thread of DNA interwoven with protein that is the gene-bearing structure of eukaryotic cells (3); microscopic structure within cells that carries the molecule deoxyribonucleic acid (DNA)—the hereditary material that influences the development and characteristics of each organism.

clone A group of genetically identical cells or organisms produced asexually from a common ancestor; all cells in the clone have the same genetic material and are exact copies of the original (1).

cloning The process of asexually producing a group of cells (clones), all genetically identical, from a single ancestor; in recombinant-DNA technology, the use of various procedures to produce multiple copies of a single gene or segment of DNA is referred to as 'cloning DNA' (4).

containment (physical) Measures designed to prevent or minimise the escape of recombinant organisms (+2).

cytobiology The biology of cells.

cytogenetics The study of inheritance in relation to the structure and function of cells (+5).

cytoplasm The part of the cell that lies between the membrane and the nucleus. It contains a variety of subcellular structures that participate in the cell's functions (4).

DNA (deoxyribonucleic acid) The self-replicating molecule that is the carrier of genetic information in nearly all living organisms. Every inherited characteristic has its origin somewhere in the code of each individual's DNA (1).

double helix The name given to the structure of the DNA molecule, which is composed of two complementary strands which lie alongside and twine around each other, joined by cross-linkages between base pairs—two nucleotide bases on different strands of the nucleic acid molecule that bond together (+2).

ecologically sustainable development (ESD) Types of social and economic development which sustain the natural environment and promote social equity.

ecosystem A community of life forms, interacting with each other, together with the environment in which they live and with which they also interact (1).

enzyme Any one of a large class of proteins occurring in organisms which act as catalysts, that is, make it possible for chemical reactions to occur and to occur sufficiently rapidly to meet the organism's needs (4).

***Escherichia coli* (*E.coli*)** A bacterial species that inhabits the intestinal tract of most vertebrates and on which much genetic work has been done; some strains are pathogenic to humans and other animals; many non-pathogenic strains are used experimentally as hosts for recombinant-DNA (2).

eugenics A strategy of trying to orchestrate human evolution through programmes aimed at encouraging the transmission of 'desirable' traits and discouraging the transmission of 'undesirable' ones (3).

feedback loop A flow of information back to its origin. A circular causal process in which a system's output is returned to its input. An ecosystem's feedback loops are the pathways along which nutrients are continually recycled (for more detail see Capra 1997).

gene The basic physical and functional unit of heredity that is transmitted from one generation to the next and can be transcribed into a polypeptide or protein (3); a segment of chromosome; some genes direct the synthesis of proteins, while others have regulatory functions.

gene mapping Determination of the relative positions of genes on a chromosome.

gene therapy The application of genetic engineering techniques to alter or replace defective genes. The techniques are still at the experimental stage (+5).

genetic determinism The theory that the phenotype (the detectable characteristics associated with a particular genotype) is an innate and essentially unchangeable expression of the genotype (+2). Genetic determinism (or essentialism—see chapter 9) is now also being referred to as *geneticisation*—'an ongoing process by which differences between individuals are reduced to their DNA codes, with most disorders, behaviors and physiological variations defined, at least in part, as genetic in origin' (Lippman 1991 cited in Hubbard and Wald 1997, p. 187). A more general statement, based upon a critical perspective, is this: 'The new biotechnology based upon genetic engineering makes the assumption that each specific feature of an organism is encoded in one or a few specific, stable genes, so that the transfer of these genes results in the transfer of a discrete feature. This extreme form of genetic reductionism has already been rejected by the majority of biologists and many other members of the intellectual community because it fails to take into account the complex interactions between genes and their cellular, extra-cellular and external environments that are involved in the development of all traits.' (see Ho, 1998; Third World Network 1995, p. 10)

genetic engineering The technique of altering the genetic makeup of cells or individual organisms by deliberately inserting, removing or altering individual genes (+3).

genetic monitoring Tests that examine hereditary molecules for early indications of genetic damage or disease (3).

genetics The branch of biology concerned with the study of heredity and variation (+5).

genetic screening The process by which individuals are asked to undergo genetic testing; the process of systematically scanning individual genotypes for possible hereditary defects or abnormalities (3).

genetic testing Analysis of human DNA, RNA, chromosomes, or gene products for early indications/ of genetic damage or disease.

genome The entire genetic endowment of an organism or individual (3).

genomics The study of genomes, which includes genome mapping, gene sequencing and gene function.

genotype An individual's genetic makeup underlying a specific trait or constellation of traits (3).

germ cell Reproductive cell.

glanders A contagious disease of horses, mules, etc., communicable to humans, due to micro-organism (*Bacillus mallei*), and characterised by swellings beneath the jaw and a profuse mucous discharge from the nostrils.

GMO Genetically modified organism (see genetic engineering).

Guthrie test A biochemical test, not one based on molecular (genetic) technology.

hybrid A molecule, cell or organism produced by combining the genetic material of genetically dissimilar organisms; traditionally, hybrids were produced by interbreeding whole animals or plants; cell fusion technology and transgenic engineering are innovations in hybridisation (2).

hybridoma A 'hybrid myeloma'; a cell produced by the fusion of an antibody-producing cell (lymphocyte) with a cancer cell (myeloma) (2).

hybridoma technology The technology of fusing antibody-producing cells with tumour cells to produce in hybridomas which proliferate continuously and produce monoclonal antibodies (2).

immunology The study of phenomena related to the body's response to antigenic challenge (that is, immunity, sensitivity, and allergy).

intellectual property rights Laws that grant monopoly rights to those who create ideas or knowledge. They are intended to protect inventors against losing control of their ideas or the creations of their knowledge. There are five major forms of intellectual property: patents, plant breeder's rights, copyright, trademarks, and trade secrets. All operate by exclusion, and grant temporary monopolies which prevent others from making or using the creation. Intellectual property legislation is national, although most countries adhere to international conventions governing intellectual property, and all members of the World Trade Organisation must have intellectual property covering living organisms. Patents and plant breeder's rights are the forms most relevant to living organisms.

in vitro Biological reactions occurring outside of the living organism, in test tubes or any laboratory containers—all of which are artificial systems. Latin term meaning 'in glass'.

in vitro fertilisation **(IVF)** The fertilisation of an egg cell by sperm on a glass dish (2); fertilisation carried out in the laboratory, outside a woman's body (1).

Janus-faced two-sided; Janus was an ancient Italian god—the guardian of doors and gates, and who was represented with faces on the front and back of the head.

messenger RNA An RNA molecule that has been transcribed from a gene-bearing DNA molecule and will later be translated into a protein (3).

metachronical DNA

microbiology The study of living organisms that can only be seen under a microscope.

microorganism (microbe) An organism that can only be seen under a microscope.

molecular biology The study of the structure and function of large molecules associated with living organisms, in particular proteins and the nucleic acids DNA and RNA. (5).

molecular genetics The study of the molecular basis of gene structure and function (3).

monoclonal antibody One of a clone of antibodies produced by a hybridoma (2).

monoculture The agricultural practice of cultivating crops consisting of genetically similar organisms (3).

mutation A heritable change in a DNA molecule (3).

neo-luddite Person actively opposed to the introduction of new technology; after the Luddites, an organised band of mechanics which went about destroying machinery in the midlands and north of

England in the early nineteenth century; derived from Ned Lud, who, in the latter half of the 18th century, smashed up machinery belonging to a Leicestershire manufacturer as a protest against mechanisation (2).

nucleic acid One of a number of acidic, linear, information-bearing, biological molecules belonging to one of two chemical families—DNA or RNA (3).

nucleotide A chemical subunit composed of a sugar, phosphate and base, which makes up the nucleic acids DNA and RNA (3).

organism Any form of living entity, whether plant, animal, bacterium, protist or fungus.

patents Legal monopolies that cover a wide range of products and processes, including life forms. To be patentable, inventions must theoretically meet three basic criteria. They must be: (a) novel—they must not have been known previously to the public; (b) useful—they must do what they claim, though they need not necessarily be practical; and (c) non-obvious—they must have an 'inventive step' and constitute some notable extension of what was previously known. Patents provide exclusive legal protection to patent holders, usually for seventeen to twenty-five years. Anyone wishing to use a patented invention must receive permission from the patent holder and often must pay a royalty. In exchange for this monopoly, the patent holder must disclose or describe the invention.

pathogen A micro-organism, such as a bacterium or virus, that infects another organism to produce disease symptoms or a toxic response (1).

plant breeder's rights (plant variety rights) A form of intellectual property law that grants a plant breeder's certificate to those who breed new plant varieties. Plant breeder's rights generally contain a breeder's and research exemption that allow non-commercial use of protected varieties. In the US, recent court decisions have threatened these exemptions. There are currently two international agreements governing Plant Breeder's Rights, both of them under UPOV—the International Convention for the Protection of New Plant Varieties.

plasmid A circular, self-replicating form of DNA found in many species of bacteria that can sometimes be used as a vector to ferry recombinant genes into another species (3).

polygenic Controlled by or associated with more than one gene (3).

polypeptide A protein or portion of a protein consisting of two or more amino acids (3).

population A local group of organisms belonging to the same species and capable of interbreeding (3).

probe (molecular) A DNA or RNA molecule, labelled with a radioactive isotope, dye, or enzyme, used to locate a specific nucleotide sequence or gene on a DNA molecule.

protein A large molecule composed of one or more chains of amino acids in a specific sequence; the sequence is determined by the sequence of the nucleotides in the gene coding for the protein; proteins are the doing molecules of the cells; they are required for the structure, function, and regulation of the body's cells, tissues, and organs, and each protein has unique functions (4).

quantum theory Challenges the notion in classical physics that all physical phenomena can be reduced to the properties of hard and solid material particles. Instead, in quantum theory, these particles dissolve at the subatomic level into wave-like patterns of probabilities of interconnections. The subatomic particles have no meaning as isolated entities but can only be understood as interconnections, or correlations, between various processes of observation and measurement. Quantum physics thus demonstrates that the world cannot be fragmented into independently existing elementary units, but has to be understood as complex web of relationships between the various parts of a unified whole (Capra 1997. p. 30).

recombinant-DNA (r-DNA) A novel DNA sequence produced by artificially joining pieces of DNA from different organisms together in the laboratory (3).

reductionism the philosophical belief that phenomena or organisms are best understood by breaking them up into smaller parts (4); a belief that all phenomena find their ultimate explanation in terms of elementary physico-chemical processes and events occurring at the level of atoms and molecules (see also genetic determinism above).

reproductive technology Largely experimental procedures which involve medical intervention in the process of conception (1).

retrovirus 'Any of a family of viruses, *Retroviridae*, characterized by a unique mode of replication within the cells of their hosts. Like some other viral groups, retroviruses contain a core of the nucleic acid RNA instead of the usual DNA. Unlike other RNA viruses, retroviruses replicate as DNA rather than RNA genomes inside their hosts by means of an enzyme they carry, called reverse transcriptase. The retroviruses cause various infections in birds and mammals, including humans. Some genera also cause cancers in animals, such as feline leukemia and bovine sarcoma, and research in the mid-1980s indicated that some retroviruses may cause cancers in humans' ("Retrovirus," Microsoft® Encarta® Online Encyclopedia 2000 http://encarta.msn.com © 1997-2000).

ribosomes The particles in the cell cytoplasm on which the base sequences brought there by messenger-RNA get translated into the amino acid sequence of proteins (4). Ribosomes are the microscopic particles that synthesise proteins inside cells by reading the genetic messages that have been transcribed from DNA genes into messenger RNA. Ribosomes themselves contain some fifty different proteins.

RNA (ribonucleic acid) 'In cellular organisms, another type of genetic material, called deoxyribonucleic acid (DNA), carries the information that determines protein structure. But DNA cannot act alone and relies upon RNA to transfer this crucial information during protein synthesis (production of the proteins needed by the cell for its activities and development). Like DNA, RNA consists of a chain of chemical compounds called nucleotides. Each nucleotide is made up of a sugar molecule called ribose, a phosphate group, and one of four different nitrogen-containing compounds called bases. The four bases are adenine, guanine, uracil, and cytosine. These components are joined together in the same manner as in a deoxyribonucleic acid (DNA) molecule. RNA differs chemically from DNA in two ways: The RNA sugar molecule contains an oxygen atom not found in DNA, and RNA contains the base uracil in the place of the base thymine in DNA' ("Ribonucleic acid," Microsoft® Encarta® Online Encyclopedia 2000 http://encarta.msn.com © 1997-2000).

sequence The order of nucleotides in a nucleic acid or the order of amino acids in a protein (4).

sociology The study of society. The two principle objects of study are social facts and social processes.

somatic cell A body cell that is not destined to become an egg or a sperm (3).

species A unit of biological classification; sexually reproducing organisms are classified as belonging to the same species if they can interbreed and produce fertile offspring; the interpretation of what constitutes a species is controversial; it is especially difficult to apply the concept of species to bacteria which are not subject to reproductive isolation (2). In microbiology, variants within species are called strains.

synthesise To produce by means of synthesis (Gr. "a putting together, composition")

systems A system is a set of variables so interconnected that a change in the value of one of the variables has a determinate effect on all other variables (Mann 1983, p. 388).

Three Mile Island A nuclear reactor accident near Harrisburg, Pennsylvania in which water flow in the reactor primary circuit was interrupted, leading to increased pressure, chemical reaction and partial melting of the reactor core. Radioactivity release was minimal but the cost of decommissioning the reactor exceeded US$billion (+1).

tissue culture *In vitro* growth in nutrient medium of cells isolated from tissue.

toxin A substance, in some cases produced by disease-causing micro-organisms, that is poisonous to living organisms (2).

transgenic An organism produced by the transfer and expression of genetic material (DNA) derived from a different species (see also recombinant-DNA above).

vaccine A preparation of killed of debilitated micro-organisms—or their components or products—used to induce immunity against a disease.

Key for entries

(1) *High Tech: High Co$t*
(2) *The Bio Revolution*
(3) *Genethics*
(4) *Biopolitics*
(5) *A Concise Dictionary of Biology*

+ symbolises a shortened or sightly amended version of the cited entry.

Bibliography

Abbott, A. (1996) 'Transgenic Trials under Pressure in Germany', *Nature*, 14 March, p. 94

ABC (Australian Broadcasting Commission) (2000) 'GMO Bill Amendments Anger Green Groups', 4 December (http://www.abc.net.au/science/news/stories/s219934.htm)

Agbiotech Bulletin (1997) 'Competitiveness of Biotechnology in Europe', 5, 10, p. 8

Alibek, (2000) *Biohazard*, Arrow, London

AMRAD (1995) AMRAD and Northern Land Council Sign Agreement for Pharmaceutical Screening on Arnhem Land Plants, AMRAD News Release, 19 July

_____(1996) *Annual Report*, Melbourne

_____(1997a) AMRAD Announces Natural Products Screening Collaboration, AMRAD News Release, Melbourne, 9 January

_____(1997b) AMRAD Enters Multi-million Dollar Natural Products Screening Collaboration with Rhone-Poulenc Rorer, AMRAD News Release, Melbourne, 12 May

Anderson, J., Moore, J. and Hill, R. (1997) New Regulations for Gene Technology, Joint Press Release, DIST, Canberra

Armstrong, J. and Hooper, K. (1994) 'Nature's Medicine', *Landscape* 9, 4, pp. 12–15

Australian 10 August 1976; 19 March 1994; 8 August 1994; 23 May 1995; 19–20 October 1996; 15 October 1997

Australian Academy of Science, (1980) *Recombinant DNA: An Australian Perspective*, Australian Academy of Science, Canberra

Australian Biotechnology Association (ABA) (1993) *Australian and New Zealand Biotechnology Directory 1993*, 3rd edn, Australian Biotechnology Association Ltd, Melbourne

_____(1996a) Regulation of Genetic Engineering (http://www.aba.asn.au/leaf7.html)

_____(1996b) Biotechnology in animal agriculture (http://www.aba.asn.au/leaf4.html)

_____(1999) Submission to the Review of Business Taxation: High Growth, High-technology Industries Perspective, Australian Biotechnology Association

Australasian Biotechnology (1991) 1, 1

_____ (1996) 'Company News', 6, 6, p. 335

Australian Broadcasting Commission (ABC), *Four Corners*, 18 March 1977, 11 September 1989

Australian Conservation Foundation (ACF) (1989) Policy Statement Number 45, Genetic Engineering, 9 October, ACF, Melbourne

_____(1989) Genetic Engineering Campaign Report to ACF Council, Melbourne, December

_____(1990) Genetically Engineered Pigs Scandal Covered Up, News Media Release, 26 April

_____(1992) *Genetic Engineering: Science Before It's Time*, ACF, Melbourne

Australian Cotton Outlook (1997) January

Australian Farm Journal (1998) 8, 10, December

Australian Food Council (1996) Food Biotechnology in Australia: Issues Relating to the Application of Gene Technology, Discussion Paper, Australian Food Council, Canberra

Australian GeneEthics Network (1997) *The Gene File*, 1, Australian GeneEthics Network, Melbourne

Australian Industrial Property Organisation (1997) Australian Patents for: Microorganisms, Cell Lines, Hybridomas, Related Biological Materials and Their Use, Genetically Manipulated Organisms, AIPO, Canberra

Australian Medical Association (AMA) (1995) *Revised Code of Ethics*, Australian Medical Association, Barton, ACT

Australian Research Council (ARC) (1994) Access to Australia's Genetic Resources, Draft Report of the Australian Research Council Workshop

Australian Science and Technology Council (ASTEC) (1982) *Biotechnology in Australia*, Australian Government Publishing Service, Canberra

Avcare (1995) *Facts and Figures*, 3rd edn, Avcare (National Association for Crop Protection and Animal Health), Sydney

Avery, D. (1995) *Saving the Planet with Pesticides and Plastic: the Environmental Triumph of High-Yielding Farming*, Hudson Institute, Indianapolis

Bak, P. (1996) *How Nature Works*, Springer Verlag

Barbano, D. (1997) Bovine Growth Hormone: A Cornell University bST Fact Sheet (http://www.babybag.com/articles/bovine.htm)

Bartels, D. (1984) 'Secrecy in Biotechnology is Shortsighted', *Search*,

15, pp. 7–8

Beardsworth, A. and Keil, T. (1997) *Sociology on the Menu: An Invitation to the Study of Food and Society*, Routledge, London

Beck, U. (1992) *Risk Society: Towards a New Modernity*, Sage, London

____(1995) *Ecological Enlightenment: Essays on the Politics of the Risk Society*, Humanities Press, New Jersey

____(1996) 'Risk Society and the Provident State', in Beck, U., Giddens, A. and Lash, S. (eds) *Reflexive Modernization*, Polity Press, Cambridge

Bernstein, P. (1996) *Against the Gods: The Remarkable Story of Risk*, Wiley and Sons, New York

Biotechnology Australia (1999) Developing Australia's Biotechnology Future—Discussion Paper, Commonwealth of Australia, Canberra

Bishop, R. (1974) 'Anxiety and Readership of Health Information', *Journalism Quarterly*, 51, 1, pp. 40–6

Britz, M. (1991) 'Food Biotechnology: the Industry and the Issues', *Australasian Biotechnology*, 1, 1, pp. 30–2

Brook, S. (2001) 'Killer Virus Created in Error', *Sydney Morning Herald*, 12 January, p. 1

Brough, E., Foster, J., Norton., G. and Holdom, D. (1995) Integrating New Biotechnology in Crop Protection, Paper Presented at the 4th Pacific Rim Biotechnology Conference, Melbourne, February

Brown, D. (1997) Farmers to Fight 'Scare Stories' on Super Crops, *Daily Telegraph*, 13 January

Brown, M. (1995) Lazarus Bacteria Shows Signs of Life after 30 million Years, *Sydney Morning Herald*, 20 May, p. 1

Brumfield, J. (1980) Science Rejects DNA Fears, *The Australian*, 30 July, p. 9

Bruno, K. (1997) 'Say it Ain't Soy, Monsanto', *Multinational Monitor*, Jan–Feb, pp. 27–30

Bud, R. (1993) *The Uses of Life: A History of Biotechnology*, Cambridge University Press, New York

Bunyard, P. (1996) 'Industrial Agriculture-Driving Climate Change?', *The Ecologist*, 26, 6, pp. 290–98

Burch, D., Hulsman, K., Hindmarsh, R. and Brownlea, A. (1990) Biotechnology Policy and Industry Regulation: Some Ecological, Social and Legal Considerations, Submission to the House of Representatives Standing Committee of Industry, Science and Technology Inquiry into Genetically Modified Organisms, September

Burch, D., Lyons, K. and Lawrence, G. (1998) What Do We Mean by Green? Consumers, Agriculture and the Food Industry, Paper presented at the Critical Issues in Transnational Agrifood Systems: the Millennium and Beyond, Special Session of The Australian Sociological Association's Annual Conference, Brisbane, 3 December

Burch, D., Rickson, R. and Annels, R. (1992) 'The Growth of Agribusiness: Environmental and Social Implications of Contract Farming', in Lawrence, G., Vanclay, F. and Furze, B. (eds) *Agriculture, Environment and Society: Contemporary Issues for Australia*, Macmillan, Melbourne

Burch, D., Rickson, R. and Lawrence, G. (eds) (1996) *Globalization and Agrifood Restructuring: Perspectives from the Australasia Region*, Avebury, London

Bureau of Rural Resources (1991) Biotechnology in Australia: Perspectives and Issues for Animal Production, Working Paper WP/16/91, Bureau of Rural Resources, Canberra

Burnstein, S. (1997) 'Organics Under Siege', *Food and Water Journal*, Winter, pp. 20–1

Busch, L. (1991) Biotechnology: Consumer Concerns About Risks and Values', *Food Technology*, April, pp. 96–101

——, Lacy, W., Burkhardt, J. and Lacy, L. (1991) *Plants, Power and Profit: Social, Economic and Ethical Consequences of the New Biotechnologies*, Basil Blackwell, Massachusetts

Butler, D. (1997) Bioethics Needs Better Input From Public', *Nature* 389, 6653, p. 775

Cabinet Office (1999) 'Genetic Modification Issues: Overview of Regulatory and Advisory System'. www.gm-info.gov.uk/1999/overview.htm

Campbell, H. (2000) The Future of New Zealand's Food Exports, in Prebble, R. (ed.) *Designer Genes: the New Zealand Guide to the Issues, Facts and Theories About Genetic Engineering*, Dark Horse, Wellington

Campbell, H. and Coombs, B. (1999) '"Green Protectionism" and Organic Food Exporting from New Zealand: Crisis Experiments in the Breakdown of Fordist Trade and Agricultural Policies', *Rural Sociology* 64, 2: 302-319

Capra, F. (1997) *The Web of Life: A New Synthesis of Mind and Matter*, Flamingo, London

Carey, A. (1995) *Taking the Risk Out of Democracy*, University of New

South Wales Press, Sydney

Carey, J. (1989) *Communication as Culture. Essays on Media and Society,* Routledge, New York

CC (Cotton Conference) (1996) *Cotton On To The Future: Proceedings of The Eighth Australian Cotton Conference*, Conrad Jupiters, Gold Coast, 14–16 August

Charlesworth, M., Farrall, L., Stokes, T. and Turnbull, D. (1989) *Life Among the Scientists*, Oxford University Press, Melbourne

Clark, N. (1997) 'Panic Ecology: Nature in the Age of Superconductivity', *Theory, Culture and Society*, 14, 1, pp. 77–96

Collis, B. (1997) Reinventing Aussie Farming, *Canberra Times*, 16 September, p. 14

Connor, S. (1997) Scientists Find Gene For Intelligence, *London Sunday Times*, 19 October

CornerHouse (1997) *No Patents on Life!*, The CornerHouse, Dorset, UK, September

Cotton Research and Development Corporation (CRDC) (1997) The Performance of INGARD Cotton in Australia 1996/97 Season, Occasional Paper, CRDC, Narrabri

Cribb, J. (1994) The Next Generation, *Australian*, 10 August, p. 9

Crick, F (1958) 'On Protein Synthesis', *Symp. Soc. Exp. Biol.*, 12, pp. 138–63

_____(1970) 'Central Dogma of Molecular Biology', *Nature*, 227, pp. 561–63

Crook, S (1999) 'Ordering risks' in D. Lupton (ed) *Risk and Sociocultural Theory*, Cambridge University Press, Cambridge

Crotty, P. (1995) 'The Role of the Social Sciences in Public Health Nutrition: A Discussion Paper' in Germov, J. and Williams, L. (eds) Proceedings of *The Australian Sociological Association* Conference, December 1995, Laurijon, University of Newcastle

CSIRO (1981) Biotechnology For Australia: Report to the Executive of CSIRO, June

Daniels, K. and Murnane, M. (1980) *Uphill All The Way: A Documentary History of Women in Australia,* University of Queensland Press, Brisbane

Danks, D. (1993) 'Whither Genetic Services?', *The Medical Journal of Australia*, 159, pp. 221–22

Darby, A. and Metherell, M. (2000) 'GM Crops are Pests, Tasmania Declares', *Sydney Morning Herald*, 21 July (htpp://www.smh.com.au/news/0007/21/national/national1.html)

Davis, A. (1984) Ethical Issues in Prenatal Diagnosis, *British Medical Journal*, 288

Dawkins, K. (1997) Institute for Agriculture and Trade Policy, correspondence to US Secretary of State Madeleine Albright, 30 June

Dawkins, R. (1986) *The Blind Watchmaker*, Longman Scientific and Technical, Harlow

Dessaix, R. (1994) Another Country, *24 Hours*, May 1994, p. 148

Douglas, M. (1966) *Purity and Danger*, Routledge and Kegan Paul, London

Dow, S. (1997) Gene Patent Decision Under Fire, *The Age*, 10 June

Draper, E. (1991) *Risky Business: Genetic Testing and Exclusionary Practices in the Hazardous Workplace*, Cambridge University Press, New York

Dredge, R. (1990) Case of the Super Pig that Went to Market, *The Age*, 2 May

Dunwoody, S. (1993) *Reconstructing Science for Public Consumption: Journalism as Science Education*, Deakin University, Victoria

Duff, J. (1994) 'Eating Away at the Public Health: a Political Economy of Nutrition' in Waddell, C. and Petersen, A.R. (eds) *Just Health: Inequality in Illness, Care and Prevention*, Churchill Livingstone, Melbourne

_____(1999) 'Setting the Menu: Dietary Guidelines, Corporate Interests, and Nutrition Policy' in Germov, J. and Williams, L. (eds) *A Sociology of Food and Nutrition: The Social Appetite*, Oxford University Press, Melbourne

East Asia Analytical Unit (1994) Subsistence to Supermarket: Food and Agricultural Transformation in South-East Asia, Department of Foreign Affairs and Trade, Australian Government Publishing Service, Canberra

Eckersly, R. (1981) Biotechnology: It Takes Time to Commercialise, *Sydney Morning Herald*, 19 November, p. 7

Economist, 27 June 1987, pp. 87–93; 19 March 1994

Eder P. (1994) 'Privacy on Parade', *The Futurist*, 28, pp. 38–42

Elmer-Dewitt, P. (1994) The Genetic Revolution, *Time*, 9, 3, pp. 20–7

ENN (2000) 'No quick fix seen for U.S. biotech food chaos' 24 October 2000, *Environmental News Network* (http://www.enn.com/news/wire-stories/2000/10/10242000/ reu_biofood_39515.asp)

ENNDirect (2000) 'Gerber Parent Novartis Eliminates Genetically Engineered Food From All Products' 4 August 2000 ENN Direct (http://www.enn.com/direct/displayrelease.asp?id=1865)

Ennor, A. (Secretary, Department of Science) (1977) Correspondence to Dr C. Evans (Acting Director-General of Health, Department of Health) 24 June

Enright, J. (ed.) (1983) *The Oxford Book of Death,* Oxford University Press, Oxford

Equal Employment Opportunity Commission (1995) *Compliance Manual,* vol. 2, Section 902, Order 915.002, pp. 902–45; cited in Rothenberg, K. *et al.* (1997) 'Genetic Information and the Workplace: Legislative Approaches and Policy Challenges', *Science,* 275, pp. 1755–7

European Federation of Biotechnology (EFB) (1994) Biotechnology in Food and Drinks (http://www.kluyver.stm.tudelft.nl/tjpb/eng2.htm)

Evans, C. (Acting Director-General of Health, Department of Health) (1977) Correspondence to A. Ennor, Secretary, Department of Science, 1 July

Ewing, T. (1994) Patent on Genes Challenged, *The Age,* 9 December, p. 3

Farrands, J. (Secretary Department of Science and the Environment) (1979) Summary of Meeting on Recombinant-DNA (10 August), attached paper to letter dated 21 September 1979 from J. Waterman, Executive Assistant, Dept. of Science and the Environment, to Dr Evans, Australian Academy of Science

Fitt, G., Mares, C. and Llewellyn, D. (1994) 'Field Evaluation and Potential Ecological Impact of Transgenic Cotton (*Gossypium hirsutum*) in Australia', *Biocontrol Science and Technology,* 4, pp. 535–48

Forrester, N. (1994) 'Resistance Management Options for Conventional *Bacillus thuringiensis* and Transgenic Plants in Australian Summer Field Crops', *Biocontrol Science and Technology,* 4, pp. 549–53

_____(1997) Principal Research Scientist, NSW Agriculture and Australian Cotton Research Institute, pers. comm., January–June

Forrester, N. and Constable, G. (1997), respectively, Principal Research Scientist, NSW Agriculture and Australian Cotton Research Institute; Director, Co-operative Research Centre for Sustainable Cotton, pers comm., 26 June

Foster, J. and Ghonim, S. (1995) 'Prospects for the Commercial Use of Transgenic Plants: Attitudes of Crop Protection Professionals', *Australian Agribusiness Review,* 3, 2, December, pp. 73–94

Fourmile, H. (1995) 'Intellectual Property Rights Need Recognition', *Land Rights Queensland,* October–November, p. 7

____(1996) 'Protecting Indigenous Property Rights in Biodiversity', *Current Affairs Bulletin*, February–March, pp. 36–41

____(1997), pers. comm., 12 August

Fowler, R. (1991) *Language in the News: Discourse and Ideology in the Press*, Routledge, London

Friedman, S. (1986) A Case of Benign Neglect: Coverage of Three Mile Island Before the Accident, in Friedman S., Dunwoody, S. and Rogers, C. (eds) *Scientists and Journalists: Reporting Science as News*, Macmillan, New York

Fuller, K. (DST) (1983) Internal Memo to the Minister, 25 May

Gans, L. (1997) Judged by Your Genes, *The Courier-Mail*, 8 August, p. 18

Geisler, C. and Lyson, T. (1993) 'Bio-technology: Rural Policy Implications of Bovine Growth Hormone Adoption in the USA', in Harper, S. (ed.) *The Greening of Rural Policy: International Perspectives*, Belhaven, London

GMAC (Genetic Manipulation Advisory Committee) (2000) 'GR-9 General (commercial) release of Roundup Ready and Roundup Ready/INGARD cotton' (http://www.health.gov.au/tga/gene/gmac/gr09.htm)

Gene File, Genetic Engineering Action Update (1994) The Australian Genethics Network, Australian Conservation Foundation, Melbourne

Gene Technology Information Unit (GTIU) (1995) Gene Technology at Work, *Today's Technology*, 3

Genetic Resources Action International (GRAIN) (1996) The Biotech Battle Over the Golden Crop, *Seedling*, 13, 3, pp. 23–32

George, S. (1976) *How the Other Half Dies*, Penguin, England

Giddens, A. (1991) *Modernity and Self-Identity*, Polity Press, Cambridge

Gilchrist, G. (1995) Gene Therapy May be Used to Treat Brain Disorders, *Sydney Morning Herald*, 28 October, p. 3

Glanzig, A.(1996) Magnificent Seven is a Natural Act, *Australian*, 5 June, p. 18

Godard, B. and Verhoef, M. (1996) 'DNA Banking: Current and Ideal Practices', in Knoppers, B., Caulfield, T. and Kinsella, T. (eds) *Legal Rights and Human Genetic Material*, Emond Montgomery Publications, Toronto

Goleby, R. (Assistant Secretary Policy Division Department of Science) (1978) Internal Correspondence to the Secretary, Dept of Science (77/537)

Goodell, R. (1986) 'How to Kill a Controversy: The Case of Recombinant DNA', in Friedman, S., Dunwoody, S. and Rogers, C. (eds) *Scientists and Journalists: Reporting Science as News*, Macmillan, New York

Gottweis, H. (1995) 'German Politics of Genetic engineering and Its Deconstruction', *Social Studies of Science* 25, pp. 195–235

Gould, F., Martinez-Ramirez, A., Anderson, A., Ferre, J., Silva, F. and Moar, W. (1992) 'Broad-spectrum Resistance to *Bacillus thuringiensis* Toxins in *Heliothis virescens*', *Proceedings of the National Academy of Sciences, USA* 89, pp. 7989–90

Gray, I., Stehlik, D. and Lawrence, G. (1997) Economic Management, Social Contradiction and Climatic Disaster: Drought Policy in the 1990s, Paper Presented at the Rural Australia 2000 Conference, Charles Sturt University, Wagga Wagga, 2–4 July

Gray, I. and Lawrence, G. (2001) *A Future for Regional Australia: Escaping Global Misfortune*, Cambridge University Press, Cambridge

Green, R. (DST) (1984) Interdepartmental correspondence, 19 September

Greens in the European Parliament n.d. Opposition to European Patent No. 0112149 (Howard Florey Institute of Experimental Physiology and Medicine, University of Melbourne, Australia), Submission in the European Patent Office

Griffiths, A., Miller, J., Suzuki, D., Lewontin, R. and Gelbart, W. (1996) *An Introduction to Genetic Analysis*, 6th edn, WH Freeman and Company, New York

Grifo, F., Newman, D., Fairfield, A., Bhattacharya, B. and Gruppenhoff, J. (1997) 'The Origins of Prescription Drugs', in Grifo, F. and Rosenthal, J. (eds) *Biodiversity and Human Health*, Island Press, Washington, DC

Grobe, D., Douthitt, R. and Zepeda, L. (1999) 'A Model of Consumers' Risk Perceptions Towards Recombinant Bovine Growth Hormone (rhGH): The Impact of Risk Characteristics'. *Risk Analysis* 19, 4, pp. 661-73

Ground Cover (2000a) 'Farmers on Gene Technology' Issue 30, Ground Cover (http://www.grdc.com.au/growers/gc/gc30/biotechnology.htm)

_____(2000b) 'What bothers consumers about gene technology' Issue 30, Ground Cover (http://www.grdc.com.au/growers/gc/gc30/biotechnology.htm)

Hall, J., Lee M., Newman B., Morrow J., Anderson L., Huey B. and King

M-C. (1990) 'Linkage of Early-onset Familial Breast Cancer to Chromosome 17q21', *Science*, 250, pp. 1684–89

Hansard (1977) Senate, 23 February, p. 283

Harper, S (ed.) (1993) *The Greening of Rural Policy*, Belhaven Press, London

Haynes, R. (1991) *High Tech, High Co$t*, Pan Macmillan Australia, Sydney

——(1994) *From Faust to Strangelove: Representations of the Scientist in Western Literature*, John Hopkins University Press, Baltimore

Hazen, R. and Trefil, J. (1990) *Science Matters. Achieving Scientific Literacy*, Doubleday, New York

Healy, M. (1991) *Strategic Technologies for Maximising the Competitiveness of Australia's Agriculture-Based Exports*, IP/2/91, Bureau of Rural Resources, Canberra

Henningham, J. (1995) Who are Australia's Science Journalists? *Search*, 26, 3, pp. 89–94

Herman, E. and Chomsky, N. (1994) *Manufacturing Consent: The Political Economy of the Mass Media*, Vintage, London

Hindmarsh, R. (1992a) 'CSIRO's Genetic Engineering Exhibition: Public Acceptance or Public Awareness?', *Search*, 23, 7, pp. 212–3

——(1992b) 'Agricultural Biotechnologies: Ecosocial Concerns for a Sustainable Agriculture', in Lawrence, G., Vanclay, F. and Furze, B. (eds) *Agriculture, Environment and Society: Contemporary Issues for Australia*, Macmillan, Melbourne

——(1994) Power Relations, Social-Ecocentrism, and Genetic Engineering: Agro-biotechnology in the Australian Context (unpublished doctoral thesis Griffith University)

——(1996) 'Bio-Policy Translation in the Public Terrain', in Lawrence, G., Lyons, K. and Momtaz, S. (eds) *Social Change in Rural Australia: Perspectives from the Social Sciences*, Central Queensland University, Rockhampton

——(1998) 'Globalisation and Gene-Biotechnology: From the Centre to the Semi-Periphery', in Burch, D., Rickson, R. and Lawrence, G. (eds) *Australasian Food and Farming in a Globalised Economy: Recent Developments and Future Prospects*, Monash University, Melbourne

——(1999) 'Consolidating Control: PVR, Genes and Seeds', *Australian Journal of Political Economy*, 44: 58-78.

——(2000) 'The Problems of Genetic Engineering', *Peace Review*, 12, 4, pp. 541-47

_____, Burch, D. and Hulsman, K. (1991) 'Agrobiotechnology in Australia: Issues of Control, Collaboration and Sustainability', *Prometheus* 9, 2, pp. 221–48

Hindmarsh, R. and Hulsman, K. (1992) 'Gene Technology: the Threat or the Glory?', *New Scientist,* (Australian supplement) 25 April, p. 4

_____ (2000) 'Ethical Practice for Biotechnology' *Australasian Science*, 21, 5, pp. 17–18.

Hindmarsh, R., Lawrence, G. and Norton, J. (1995) 'Manipulating Genes or Public Opinion?' *Search*, 26, pp. 117–21

Ho, M-W (1998) *Genetic Engineering: Dream or Nightmare?* Third World Network, Penang, Malaysia

_____ (2000) 'The End of Bad Science and Beginning Again with Life', Conference on The Limit of Natural Selection, French Senate, March 18 (http://www.i-sis.org/paris.shtml)

Hoad, B. (1977) 'The Dangers of Life from a Test-tube: The New Frontier in Australian Science', *Bulletin*, 17 September, pp. 38–44

Hobbelink, H. (1991) *Biotechnology and the Future of World Agriculture*, Zed Books, London

Holman, H. and Dutton, D. (1978) 'A Case for Public Participation in Science Policy Formation and Practice', *Southern California Law Review* 51, pp. 1505–34

Holub, M. (1990) *The Dimension of the Present Moment: And Other Essays* (ed. and trans. David Young), Faber and Faber, London

_____(1992) 'Jumping to Conclusions', *Island*, 53, pp. 16–21

_____(1993) 'This Long Disease', *Island*, 54, pp. 4–6

Horin, A. (1976) Genetics—the Myth Becomes Reality, *The National Times*, 11–16 October, pp. 16–17

Hornig Priest, S. (1995) 'Information Equity, Public Understanding of Science, and the Biotechnology Debate', *Journal of Communication*, 45, 1, pp. 39–54

Howarth, F. (1984) Biotechnology: Revolutionary or Evolutionary?, unpub. MSc thesis, NSW University

Hoy, A. (1996a) Groups Round Up Support for "Gene Bean" Supermarket Warning, *Sydney Morning Herald*, 16 December, p. 3

_____(1996b) Awassi Project Aims to Put Home-grown Roquefort on Menu, *Sydney Morning Herald*, 16 April, p. 9

_____(1996c) Handless Robot's Green Thumbs-up for the Forests, *Sydney Morning Herald*, 5 November, p. 4

Hubbard, R. (1988) 'Eugenics: New Tools, Old Ideas', in Hoffmann Baruch, E., D'Adamo, A. and Seager, T. (eds) *Embryos, Ethics, and Women's Rights: Exploring the New Reproductive Technology*, The Haworth Press, New York

____(1995) 'Human Nature', in Shiva, V. and Moser, I. (eds) *Biopolitics*, Zed Books, London

____and Wald, E. (1997) *Exploding the Gene Myth*, Beacon Press, Boston

Hudson, K., Rothenberg, K., Andrews L., Ellis Kahn, M. and Collins, F. (1995) 'Genetic Discrimination and Health Insurance: An Urgent Need for Reform', *Science*, 270, pp. 391–3

Hughes, S. (1986) A Good Test but Where Can You Take It?, *Guardian*, 7 July

Hutton, P. (1997) USEPA Biopesticide and Pollution Prevention Division pers comm., 17 January

Huxley, A. (1977) *Brave New World*, Grafton Books, London

Hynes, H. (1989) 'Biotechnology in Agriculture: An Analysis of Selected Technologies and Policy in the United States', *Reproductive and Genetic Engineering*, 2, 1, pp. 39–49

INRA (Europe) – ECOSA (2000) 'Eurobarometer 52.1 The Europeans and Biotechnology' (http://www.europa.eu.int/comm/research/eurobarometer-en.pdf)

International Agricultural Development (1996) Organic Focus: Expanding Supply and Demand, 16, 6, p. 23

Invetech (1989) Opportunities in Biotechnology for Australian Industry: an Overview, Invetech Operations, Victoria

IOGTR (Interim Office of the Gene Technology Regulator) (2000a) Submission to the Community Affairs References Inquiry into the Gene Technology Bill 2000, ACT

____(2000b) 'Explanatory Guide to the Draft Commonwealth Gene Technology Regulations 2000'. (www.health.gov.au.tga.genetech.htm)

____(2000c) 'Audit of Monsanto Australia Ltd: processes for the conduct of field trials in accordance with GMAC recommendations'. *IOGTR Information Bulletin 3 (*www.health.gov.au.tga.genetech.htm)

____(2000d) 'Audit of Aventis Cropscience Pty Ltd: conduct of field trials in accordance with GMAC recommendations'. *IOGTR Information Bulletin 5* (www.health.gov.au.tga.genetech.htm)

ISAAA (International Agricultural Service for the Acquisition of

Agri-biotech Applications) (1997) Progressing Public-Private Sector Partnerships in International Agricultural Research and Development. Brief no. 4–1997

Jackson, G. (2000) 'Suzuki's *Altered Genes*', *The New Australian*, 158, 17–30 July (http://www.newaus.com.au/news158suzuki.html)

Jackson, W. (1984) 'A New Threat to Agriculture and a Window of Opportunity', *The Ecologist*, 14, pp. 119–24

James, C. and Krattiger, A. (1996) Global Review of the Field Testing and Commercialization of Transgenic Plants: 1986 to 1995, International Service for the Acquisition of Agri-biotech Applications, New York

Jennings, P. (1994) 'Biotechnology and the Biological Industries in the 21st Century', in Eckersley, R. and Jeans, K. (eds) *Challenge to Change: Australia in 2020*, CSIRO Publications, Australia

Joseph, R. and Johnston, R. (1985) 'Market Failure and Government Support for Science and Technology: Economic Theory Versus Political Practice', *Prometheus* 3, 1, pp. 138–55

Joyce, C. (1990) 'US Approves Trials with Gene Therapy', *New Scientist*, 11 August, p. 7

Judson, H. (1979) *The Eighth Day of Creation: Makers of the Revolution in Biology*, Simon and Schuster, New York

Kamaldeen, S. and Powell, A. (2000) 'Public Perceptions of Biotechnology' Food Safety Network Technical Report #17, Department of Plant Agriculture, University of Guelph (http://www.plant.uoguelph.ca/safefood/gmo/public-perceptions-biotech-aug00.htm)

Katz Rothman, B. (1986) *The Tentative Pregnancy*, Penguin, New York

Kauffman, S. (1993) *Origins of Order; Self-organization and Selection in Evolution*, Oxford University Press, New York

____(1996) Investigations, Technical Report #96-08-072, Santa Fe Institute, Santa Fe, New Mexico, USA

Kawar, A. (1989) 'Issue Definition, Democratic Participation and Genetic Engineering', *Policy Studies Journal*, 17, 4, pp. 719–44

Kidwell, M. and Lisch, D. (1997) 'Transposable Elements as Sources of Variation in Animals and Plants', *Proceedings of the National Academy of Sciences USA* 94, pp. 7704–11

Kingsnorth, P. (1998) Bovine Growth Hormones, *The Ecologist* 28, 5, Sept/Oct, pp. 266-269

Kloppenburg, J. (1988) *First the Seed: The Political Economy of Plant Biotechnology*, Cambridge University Press, Cambridge

____(1991) 'Alternative Agriculture and the New Biotechnologies', *Science as Culture*, 13, pp. 482–505

Kloppenburg, J. and Burrows, B. (1996) 'Biotechnology to the Rescue? Twelve Reasons Why Biotechnology is Incompatible with Sustainable Agriculture', *The Ecologist*, 26, 2, March–April, pp. 61–7

Knoppers, B., Caulfield, T. and Kinsella, T. (eds) (1996) *Legal Rights and Human Genetic Material*, Emond Montgomery, Toronto

Kretzschmar, H, Stowring, L, Westaway, D., Stubblebine, W., Prusiner, S. and Dearmond, S. (1986) 'Molecular Cloning of a Human Prion Protein cDNA', *DNA*, 5, 4, pp. 315–24

Krimsky, S. (1982) *Genetic Alchemy: The Social History of the Recombinant-DNA Controversy*, MIT Press, Cambridge, Massachusetts

____(1991) *Biotechnics and Society: The Rise of Industrial Genetics*, Praeger, New York

____and Wrubel, R. (1996) *Agricultural Biotechnology and the Environment: Science, Policy and Social Issues*, University of Illinois, Urbana

Landline (1997) Australian Broadcasting Corporation, 4 May

Lasker, J. and Borg, S. (1987) *In Search of Parenthood: Coping With Infertility and High-Tech Conception*, Beacon Press, Boston

Latour, B. (1987) *Science in Action: How to Follow Scientists and Engineers through Society*, Open University Press, Cambridge

Law Reform Commission of Victoria (1988) Genetic Manipulation, Discussion Paper No. 11, Law Reform Commission of Victoria, Melbourne

Lawrence, G. (1987) *Capitalism and the Countryside: The Rural Crisis in Australia*, Pluto, Sydney

____(1995) Futures for Rural Australia: From Agricultural Productivism to Community Sustainability, Inaugural Address, Central Queensland University, RSERC, Rockhampton

____and Norton, J. (1994) Industry Involvement in Australian Biotechnology: the Views of Scientists, *Australasian Biotechnology*, 4, 6, pp. 362–8

Lawrence, G. and Vanclay, F. (1992) 'Agricultural Production and Environmental Degradation in the Murray-Darling Basin', in Lawrence, G., Vanclay, F. and Furze, B. (eds) *Agriculture, Environment and Society: Contemporary Issues for Australia*, Macmillan, Melbourne

Lawrence, G., Vanclay, F. and Furze, B. (eds) (1992) *Agriculture, Environment and Society: Contemporary Issues for Australia*, Macmillan, Melbourne

Lawrence, G., Lyons, K. and Momtaz, S. (eds) (1996) *Social Change in Rural Australia*, RSERC, Central Queensland University, Rockhampton

Lederberg, J. (1991) 'Pandemic as a Natural Evolutionary Phenomenon', in Mack, A. (ed.) *In Time of Plague. The History and Social Consequences of Lethal Epidemic Disease*, New York University Press, New York

Lee, M. (1992a) Genetic Manipulation: the Threat or the Glory?, House of Representatives. Standing Committee on Industry, Science and Technology, Australian Government Publishing Service, Canberra

_____(1992b) Media Release: Inquiry into Genetically Modified Organisms, House of Representatives Standing Committee on Industry, Science and Technology, 26 March

Lester, I. (1994) *Australia's Food and Nutrition*, Australian Government Publishing Service, Canberra

Lewin, B. (1994) *Genes V*, Oxford University Press, Oxford

Lifton, R. (1987) *The Nazi Doctors*, Macmillan, London

Llewellyn, D (1997) (CSIRO Division of Plant Industry), pers. comm., 28 January; 6 June

Llewellyn, M. (1995) How the Genes Revolution Poses Human Dilemmas, *The Northern Herald*, 21 September, p. 6

Lloyd, G. (1997) Farewell To Privacy, *The Courier Mail*, 12 April, p. 26

Loge, P. (1991) Language is a Virus: Discourse and the Politics and Public Policy of Biotechnology, Paper prepared for the 12th International Meeting of the Society of Environmental Toxicology and Chemistry, Seattle, US, 6 November

Love, R. (1993) 'The Public Relations of Science, the Flying Pink Pig and the Jet-propelled Cane-toad', *Chain Reaction*, 68, pp. 21–2

Lövel, G. (1997) 'Global Change Through Invasion', *Nature*, 388, pp. 627–8

Lyons, K. and Lawrence, G. (1999) 'Alternative Knowledges, Organic Agriculture and the Biotechnology Debate', *Culture and Agriculture* 21, 2: 1-12

Lyons, K., Lockie, S. and Lawrence, G. (2000) Consuming 'Green': the Symbolic Construction of Organic Foods, Paper presented at the VIII Agri-food Research Network Conference, Tumbarumba, NSW, 30 November-2 December

McAuliffe, S. and McAuliffe, K. (1981) *Life for Sale*, Coward, McCann & Geoghegan, New York

McEwen, J. and Reilly, P. (1994) 'Stored Guthrie Cards as DNA "Banks"', *American Journal of Human Genetics* 55, pp. 196–200

Macgregor, N. (2000) Footprints of Genetic Engineering in Agriculture, in Prebble, R. (ed.) *Designer Genes: the New Zealand Guide to the Issues, Facts and Theories About Genetic Engineering*, Dark Horse Press, Wellington

McLaren, A. (1987) 'Can We Diagnose Genetic Disease in Pre-embryos?', *New Scientist*, 10 December, p. 42

McLean, G., Waterhouse, P., Evans, G. and Gibbs, M. (1997) *Commercialisation of Transgenic Crops: Risk, Benefit and Trade Considerations*, (Department of Primary Industries and Energy/Bureau of Resource Sciences) Australian Government Publishing Service, Canberra

Maddox, J. (1993) 'Wilful Public Misunderstanding of Genetics', *Nature*, 364, p. 281

Malkin, D., Li F., Strong L., Fraumeni J. Jr., Nelson C., Kim D., Kassel J., Gryka M., Bischoff F., Tainsky M. and Friend S. (1990) 'Germ line *p53* Mutations in a Familial Syndrome of Breast Cancer, Sarcomas, and Other Neoplasms', *Science*, 250, pp. 1233–8

Marsden, T. (1992) 'Exploring a Rural Sociology for the Fordist Tradition', *Sociologia Ruralis*, XXXII, 2–3, pp. 209–30

Martin, E. (ed.) (1990) *A Concise Dictionary of Biology*, Oxford University Press, London

Marx, J. (1993) 'Cell Death Studies Yield Cancer Clues', *Science*, 259, pp. 760–1

Masood, E. (1996) 'Gene Tests: Who Benefits From Risk?', *Nature*, 379, pp. 389–92

Mayeno, A. and Gleich, G. (1994) 'Eosinophilia-myalgia syndrome and tryptothan production: a cautionary tale', *Trends in Biotechnology* 12, pp. 346–52

Mazur, A. (1981) 'Media Coverage and Public Opinion on Scientific Controversies', *Journal of Communication*, 31, pp. 106–15

Mensah, R. (1996) (IPM researcher, NSW Agriculture) pers. comm.

Millis, N. (Chair, RDMC) (1984) Correspondence to Barry Jones (Minister for Science and Technology, 17 December

_____(Chair, RDMC) (1986) Correspondence to Prof D. Shanks, Vice Chancellor, Adelaide University, 9 July

Mills, R. (1990) Decision Time for Genetic Engineering, *Australian*

Rural Times, 21–27 June, p. 6

Mitcham, C. (1997) 'Justifying Public Participation In Technical Decision Making', *IEEE Technology and Society Magazine* 16, 1, pp. 40–6

Monsanto (1995) *Introducing Ingard: Some Basic Facts and Perspectives*, Monsanto Australia, Melbourne

____(1995) Roundup Ready Gene Agreement 1996, Monsanto Agreement for Roundup Ready Soybeans, Monsanto

____(nd) Results of Bollgard—Research Regarding Bollworm Infestations, Monsanto

____(1997) Effective Weed Control

Morton, O. (1997) 'First Dolly, Now Headless Tadpoles', *Science* 278, 5339, p. 798

Moser, I. (1995) 'Critical Communities and Discourses on Modern Biotechnology', in Shiva, V. and Moser, I. (eds) *Biopolitics*, Zed Books, London

Mussared, D. (1994) Yes, We'll Eat those Tomatoes, *Canberra Times*, 18 October, p. 12

Myers, G. (1990) 'The Double Helix as Icon', *Science as Culture*, 9, p. 49

National Action Plan on Breast Cancer (NAPBC) and the NIH–DOE Working Group on the Ethical, Legal, and Social Implications of Human Genome Research. (1995) Genetic Discrimination and Health Insurance: A Case Study on Breast Cancer', Bethesda, MD, 11 July 1995, Workshop (cited in Hudson, K. *et al.* (1995) 'Genetic Discrimination and Health Insurance: An Urgent Need for Reform', *Science*, 270, pp. 391–3)

National Farmers' Federation (1993) New Horizons: A Strategy for Australia's Agrifood Industries, Canberra

National Health and Medical Research Council (NHMRC) (1995) *Aspects of Privacy in Medical Research*, Australian Government Printer, Canberra

National Occupational Health and Safety Commission (1994) Control of Workplace Hazardous Substances: National Model Regulations for the Control of Workplace Hazardous Substances, Australian Government Publishing Service, Canberra

National Registration Authority (NRA) (1996a) Public Release Summary: INGARD Gene by Monsanto, National Registration Authority for Agricultural and Veterinary Chemicals, Canberra, 7 May

____(1996b) NRA Community Brief, National Registration Authority for Agricultural and Veterinary Chemicals, Canberra, 6 August

Nature (1992) 'Genetics and the Public Interest', 356, 2 April, pp. 365–6

____(1997) 'Council of Europe Urged to Ban Human Cloning', 389, 6652, p. 656

Nature Biotechnology (1996) 'Soya, Supermarkets, Sense and Segregation', 14 December, p. 1627

Neale, G. (1997) 'Frankenstein Food' Faces Supermarket Ban, *Sunday Telegraph*, 26 January, London

Nelkin, D. (1987) *Selling Science: How the Press Covers Science and Technology*, W.H. Freeman and Company, New York

____(1996) 'Genetics, God and Sacred DNA', *Society*, 33, 4, pp. 22–5

Nelkin, D and Lindee, M. (1995) *'The DNA Mystique: The Gene as a Cultural Icon'*, W.H. Freeman, New York

New York Times (1977) 24 July

Norton, J. (1999) 'Science Technology and the Risk Society: Australian Consumers' Attitudes to Genetically-Engineered Foods', unpublished thesis, Central Queensland University, Rockhampton

Norton, J., Lawrence, G., Wood, G. (1998) 'The Australian Public's Perception of Genetically-engineered Foods', *Australasian Biotechnology*, 8, 3, May/June, pp. 172-181.

Nottingham, S. (1998) *Eat Your Genes: How Genetically Modified Food is Entering Our Diet*, Australian Consumers' Association, Marrickville

Nurse, P. (1997) 'The Ends of Understanding', *Nature*, 387, p. 657

O'Brien, S. (2000) 'Which Came First, the Chicken or the Egg?',

O'Connor K. (1996) *The Privacy Implications of Genetic Testing, Information Paper Number Five*, Commonwealth of Australia, Canberra

OECD (Group of National Experts on Safety in Biotechnology) (1992) Public Information/Public Education in Biotechnology: Results of an OECD Survey, OECD, DSTI/STP/BS(92)7, Paris

Official Journal of Patents (1976) 'Trade Marks and Designs', 21 October, p. 3915

O'Neill, G. (1990a) Uproar over mutant meat, *The Age*, 28 April, p. 1

____(1990b) The Issue is the Right to Know, *The Age*, 2 May, p. 13

____(1991) 'Human DNA: Cracking the Code', *21C*, Summer 90–91, pp. 39–43

Organic Federation of Australia (2000) 'Gene Technology Bill Headline Issues', Press Release 4 December

Parliament of Australia (2000) Gene Technology Bill 2000, Bills Digest No. 11 2000-01 (http://www.aph.gov.au/library/pubs/bd/2000-01/01bd011.htm#Main)

Patterson, P. and Wilkins, L. (1991) *Media Ethics: Issues and Cases*, Wm. C. Brown Publishers, Dubuque

____(1994) 'Genetic Engineering of Crop Plants Will Enhance the Quality and Diversity of Foods', *Food Australia*, 46, pp. 379–81

Peacock, J. (CSIRO ·Division of Plant Industry) (1990) Internal Correspondence to J. Stocker (Chief Executive CSIRO), 7 June

____(1995) Gene Technology and Our Future Lifestyle, Australian Foundation for Science Lecture presented at ANZAAS '95 Congress, Newcastle, August

Phelps, R. (1997) Business is Eating Away at our Food Supply, *Canberra Times*, 16 October, p. 11

Playne, M. (1998) 'Public Perceptions of Genetic Engineering', *Australasian Biotechnology* 8, 1, pp. 39–42

Plein, L. (1991) 'Popularizing Biotechnology: The Influence of Issue Definition', *Science, Technology, and Human Values*, 16, 4, pp. 474–90

Pocock, I. (1997) GE-Irish Update, 15 August, source-email address: gentech@tribe.ping.de

Pokorski, R. (1995) 'Genetic Information and Life Insurance', *Nature*, 376, pp. 13–4

Poulter, S. (1999) 'Sainsbury's to axe GM Own Brands', P 2, 17 March 1999, *Daily Mail*, London

Prakash, C. (1997) A Commentary on the Proceedings of the National Academy of Science Paper Concerning Bt Cotton Adaptation by Insects, from *ISB NewsReport*, July (http://www.nbiap.vt.edu)

Prebble, R. (ed.) (2000) *Designer Genes: the New Zealand Guide to the Issues, Facts and Theories About Genetic Engineering*, Dark Horse, Wellington.

Priest, S. (1995) 'Information Equity, Public Understanding of Science and the Biotechnology Debate', *Journal of Communication*, 45, 1, pp. 39–54

Prusiner, S. (1997) 'Prion Diseases and the BSE Crisis', *Science*, 278, pp. 245–51

Pyke, B. (1997) (Entomologist, Cotton Research and Development Corporation) pers comm., January–July

Quade, V. (1993) 'Protecting the Essence of Being', *Human Rights*, 20, pp. 14–17

Quantum (1996) Australian Broadcasting Corporation, 15 August

Queensland Department of Primary Industries (2000) 'Organic Food Focus for Queensland' 4 July 2000 (media release)

Raymond, J. (1993) *Women As Wombs: Reproductive Technologies and the Battle Over Women's Freedom*, Harper, San Francisco

Recombinant-DNA Monitoring Committee (RDMC) (Department of Science of Technology) (1982) *Recombinant DNA Monitoring Committee Report for the Period October 1981 to October 1982*, Australian Government Publishing Service, Canberra

Reeves, T. (Chairman, Advisory Committee, CSIRO Division of Plant Industry) (1990) Internal Correspondence to J. Stocker (Chief Executive CSIRO) 23 May

Rifkin, J. (1984) *Algeny*, Penguin, Harmondsworth

____(1998) *The Biotech Century*, Tarcher/Putnam, New York

Ripe, C. (1996) Designer Genes, No Label, *Australian*, 13 December, p. 17

Rissler, J. (1997) 'Bt Cotton-Another Magic Bullet?', *Global Pesticide Campaigner*, 7, 1

Rissler, J. and Mellon, M. (1990) 'Public Access to Biotechnology Applications', *Natural Resources and Environment*, pp. 4, 3, pp. 29–31, 54–8

Roberts, L. (1990) 'L-tryptophan Puzzle Takes New Twist', *Science*, 249, 31 August, p. 988

Rose, S. and Rose, H. (1976) 'The Politics of Neurobiology: Biologism in the Service of the State', in Rose, H. and Rose, S. (eds) *The Political Economy of Science*, MacMillan, UK

Rothenberg, K., Fuller, B., Rothstein, M., Duster, T., Ellis Kahn, M., Cunningham, R., Fine, B., Hudson, K., King, M-C., Murphy, P., Swergold, G. and Collins, F. (1997) 'Genetic Information and The Workplace: Legislative Approaches and Policy Challenges', *Science,* 275, pp. 1755–7

Roush, R. (1997) Associate Professor, Department of Crop Protection, Waite Institute, University of Adelaide, pers. comm., January–July

Rowell, A. (1996) *Green Backlash: Global Subversion of the Environment Movement*, Routledge, London

Rowland, R. (1992) *Living Laboratories: Women and Reproductive Technologies*, Pan Macmillan Australia, (distributed by Spinifex Press, Melbourne)

Rural Advancement Foundation International (RAFI) (1989) Biotechnology Industry Consolidation, *RAFI Communique*,

November, pp. 1–12

____(1994) Declaring the Benefits, *RAFI Occasional Paper*, 1 (3), pp. 1–14

____(1996a) The Life Industry, *RAFI Communique*, September, pp. 1–8

____(1996b) Pharmaceutical Companies Bid for Northern Botanical Garden Collections in Attempt to Avoid the Biodiversity Convention, *RAFI Communique*, July–August, p. 10

____(1997) Bioserfdom: Technology, Intellectual Property and the Erosion of Farmers' Rights in the Industrialized World, *RAFI Communique*, March–April, pp. 1–10

Saltus, R. (1986) Biotech Firms Compete in Genetic Diagnosis, *Science*, 234, p. 1319

Saxton, M. (1988) 'Prenatal Screening and Discriminatory Attitudes About Disability', in Hoffmann Baruch, E., D'Adamo, A. and Seager, T. (eds), *Embryos, Ethics, and Women's Rights: Exploring the New Reproductive Technology*, The Haworth Press, New York

Schattschneider, E. (1960) *The Semi-Sovereign People*, Holt, Rinehart and Winston, New York

Schmeck, H. (1986) Crystal Ball is Ethically Dark, *The Age*, 3 November

Schmidt, K. (1997) 'It Was My Genes Guv', *New Scientist*, 2107, pp. 46–50

Scott-Kemmis, D., Darling, T., Johnston, R., Collyer, F. and Cliff, C. (1990) *Strategic Alliances in the Internationalisation of Australian Industry*, Australian Government Publishing Service, Canberra

Scott-Kemmis, D. and Darling, T. (1991) Biotechnology and Processed Food, *Australasian Biotechnology*, 1, 1, pp. 32–5

Shand, H. (1997) Human Nature: Agricultural Biodiversity and Farm-based Food Security, United Nations Food and Agriculture Organization, Rome

Shapin, S. (1988) 'Following Scientists Around', *Social Studies of Science* (Essay Review), 18, pp. 533–50

Shapiro, J. (1997) Genome Organization, Natural Genetic Engineering and Adaptive Mutation, *Trends in Genetics*, 13, 3, pp. 98–104

Shiva, V. (1997) 'Dr Vandana Shiva Responds', *The Ecologist* (Letter Forum), 27, 5, pp. 211–2

Shiva, V. and Moser, I. (1995) *Biopolitics*, Zed Books, London

Silver, L. (1997) *Remaking Eden: Cloning and Beyond in a Brave New World*, Avon Books, New York

Sleigh, M. (1988) 'The Future of Commercial Biotechnology', *Australian Journal of Biotechnology*, 2, 1, pp. 26–7

Smith, D. (1995) One Man's Battle in Melanoma War, *Sydney Morning Herald*, 9 December, p. 27

Solé, R., Manrubia, S., Benton, M. and Bak, P. (1997) 'Self-similarity of Extinction Statistics in the Fossil Record', *Nature*, 388, pp. 764–7

Stanworth, M. (1987) 'The Deconstruction of Motherhood', in Stanworth, M. (ed.) *Reproductive Technologies: Gender, Motherhood and Medicine*, Polity Press, Cambridge

Steinbrecher, R. (1996) 'From Green Revolution to Gene Revolution: The Environmental Risks of Genetically Engineered Crops', *The Ecologist*, *26*, 6, pp. 273–81

Stocker, J. (1990) (Chief Executive CSIRO) Internal Correspondence to T. Reeves (Chairman, Advisory Committee, CSIRO Division of Plant Industry), 4 June

____(1997) *Priority Matters*, Department of Industry, Science and Tourism, Corporate Communications Section, Canberra

Stott Despoja, N. (1997) Mapping Out the Future of Genes, *Canberra Times*, 18 November, pp. 43–5

Suzuki, D., and Knudston, P. (1988) *Genethics: The Ethics of Engineering Life*, Allen & Unwin, Sydney, Australia, originally published by Stoddart Publishing Co. Ltd, Toronto, Canada

Suzuki, D. and Levine, J. (1994) *Cracking the Code: Redesigning the Living World*, Allen & Unwin, Sydney, Australia

Sweet, M. (1995) New Study Shows Families with Heart Attack Gene More at Risk, *Sydney Morning Herald*, 16 March, p. 2

____(1997) Old Drugs to Cure Modern-day Ills, *Sydney Morning Herald*, 25 February

Swinbanks, D. and Anderson, C. (1992) Search for Contaminant in EMS Outbreak Goes Slowly, *Nature*, 358, 9 July, p. 96

Sydney Morning Herald 1 April 1993; 19 June 1995; 21 July 1995; 24 July 2000; 16 December 1995; 20 October 1997, p. 11; 23 October 2000, p. 1; 23 October 2000, p. 11

The Brisbane Institute (2000) *Biobusiness: Queensland's Competitive Strengths and Challenges in the Life (Bio) Industries*, Biofutures Interim Discussion Paper, The Brisbane Institute

The Economist, 18 November 2000

The Guardian (1999) 'USA: GE crops not better or safer' 13 October 1999, *The Guardian* avail (http://www.cpa.org.au/garchve1/975gm.html)

The Morning Bulletin (2000) 'GM labelling the toughest' p. 3, 29 July 2000 *The Morning Bulletin,* Rockhampton

Third World Network (1995) *The Need for Greater Regulation and Control of Genetic Engineering: A Statement by Scientists Concerned about Current Trends in the New Biotechnology*, Third World Network, Penang, Malaysia

____(2000) Joint NGO Statement on the Review of Article 27.3 (b) of the TRIPS Agreement, 27 November 2000 (http://www.twn-side.org.sg/)

Tiffen, R. (1989) *News and Power*, Allen & Unwin, Sydney

Time 17 January 1994; 25 April 1994

Troutwine, G. (1991) 'Genetic Engineering of L-tryptophan: Futuristic Disaster', *Trial* July, pp. 20–5

Turner, F. (1998) 'The Invented Landscape'. In R. Botzler and S. Armstrong eds. *Environmental Ethics: Divergence and Convergence*, 2nd edn, McGraw Hill, Boston

Turney, J. (1993) 'Thinking About the Human Genome Project', *Science as Culture*, 19, p. 293

TWN (Third World Network) (1995) *The Need for Greater Regulation and Control of Genetic Engineering: A Statement by Scientists Concerned about Current Trends in the New Biotechnology*, Penang, TWN, Malaysia

Uhlig, R. (1996) Gene Engineered Foods are Unsafe, Says Scientist, *Daily Telegraph*, 7 September, London

University of Melbourne Assembly Report (1979) Report on Genetic Engineering, Assembly Reports

University of Melbourne (1994) The University of Melbourne Research Report 1994

US Patent Office (1997) Patent Number 5672607, Lexis-Nexus Database, 30 September

Vanclay, F. and Lawrence, G. (1995) *The Environmental Imperative: Eco-social Concerns for Australian Agriculture*, CQU Press, Rockhampton

Vandemeer, J. (1996) *Reconstructing Biology: Genetics and Ecology in the New World Order*, John Wiley & Sons, New York

Venter, C. and Cohen, D. (1997) Genetic Code-Breakers, *Weekend Australian*, 19–20 July, p. 28

Verkerk, R. (1997) IPM researcher, Imperial College of Science, Technology and Medicine, Silwood Park, UK, pers. comm.' July

____and Wright D. (1997) 'Field-based Studies on Diamondback Moth Tritrophic System in Cameron Highlands, Malaysia', *International Journal of Pest Management*, 43, pp. 27–33

Youngberg, G., Schaller, N. and Merrigan, K. (1993) The Sustainable Agriculture Policy Agenda in the United States: Politics and Prospects, in Allen, P. (ed.) *Food for the Future: Conditions and Contradictions of Sustainability*, John Wiley, New York

Verhoef, M., Lewkonia, R. and Kinsella, D. (1996) 'Ethical Implications of Current Practices in Human DNA Banking in Canada', in Knoppers, B., Caulfield, T. and Kinselle, T. (eds) *Legal Rights and Human Genetic Material*, Edmond Montgomery Publication, Toronto

Vines, G. (1986) 'Test-tube Pioneer Fears Rise of Eugenics', *New Scientist*, 9 October, p. 17

Wagner, W., Torgerson, H., Einsiedel, E., Jelsoe, E., Fredickson, H., Lassen, J., Rusanen, T., Boy, D., de Cheveigne, S., Hampel, J., Stathopoulou, A., Allansdotottir, A., Midden, C., Nielsen, T., Przestalski, A., Twardowski, T., Fjaestad, B., Olsson, S., Olofsson, A., Gaskell, G., Durant, J., Bauer, M. and Liakopoulos, M. (1997) 'Europe Ambivalent on Biotechnology', *Nature*, 387, pp. 845–7

Waldrop, M. (1992) *Complexity: The Emerging Science at the Edge of Order and Chaos*, Simon and Schuster, New York

Ward, I. (1995) *Politics of the Media*, MacMillan, Melbourne

Ward, S. (1977) Correspondence to RAFI, 13 June

Webster, G. and Goodwin, B. (1996) *Form and Transformation: Generative and Relational Principles in Biology*, Cambridge University Press, Cambridge

Weekend Australian 15-16 May 1999, p. 24

Weinberger, J. (1989) *New Atlantis and The Great Instauration Bacon*, rev. edn, Harlan Davidson, Illinois

Wells, H.G. (1975) *The Island of Dr. Moreau*, Pan, London

Wells, J. (1995) Genetic Testing May Unearth Gold, *Sydney Morning Herald*, 9 October, p. 44

Wheale, P. and McNally, R. (1988) *Genetic Engineering: Catastrophe or Utopia?*, Harvester, Wheatsheaf

_____(1990) *The Bio-Revolution: Cornucopia or Pandora's Box*, Pluto Press, London

Wilkie, T. (1995) Genetic Link with Petty Crime Claimed, *Sydney Morning Herald*, 16 February, p. 29

Wilkie, T. (1996) 'Genes "R" Us', in Robertson, G., Marsh, M., Tickner, L., Bird, J., Curtis, B. and Putnam, T. (eds) *FutureNatural: Nature, Science, Culture*, Routledge, London

Williams, R. (1989) *The Science Show*, ABC Radio, 4 February

Wills, P. R. (1989) 'Genetic Information and the Determination of Functional Organisation in Biological Systems', *Systems Research*, 6, pp. 219–26

_____(1994) 'Correcting Evolution: Biotechnology's Unfortunate Agenda', *Revue Internationale de Syste'mique*, 8, 4–5, pp. 455–68

_____(1996) 'Transgenic Animals and Prion Diseases: Hypotheses, Risks, Regulations and Policies', *New Zealand Veterinary Journal*, 44, pp. 33–6

Woodford, J. (2000) 'Crop Target 2005: A Million Hectares', *Sydney Morning Herald*, 24 July

Wooster, R., Neuhausen S., Mangion J., Quirk Y., Ford D., Collins N., Nguyen, K. and Seal S. (1994) 'Localization of a breast cancer susceptibility gene, *BRCA2*, to chromosome 13q12–13', *Science*, 265, pp. 2088–90

Working Group on Sustainable Agriculture (1991) Sustainable Agriculture, CSIRO, Melbourne

Wright, S. (1986) 'Molecular Biology or Molecular Politics? The Production of Scientific Consensus on the Hazards of Recombinant-DNA Technology', *Social Studies of Science*, 16, pp. 593–620

_____(1993) 'The Social Warp of Science: Writing the History of Genetic Engineering Policy', *Science, Technology and Human Values*, 18, 1, pp. 79–101

Wuensche, A. and Lesser, M. (1992) *The Global Dynamics of Cellular Automata*, Addison-Wesley, Reading, Massachusetts

Wynne, B. (1983) 'Redefining the Issues of Risk and Public Acceptance', *Futures*, February, pp. 13–32

_____(1991) 'Knowledges in Context', *Science, Technology and Human Values*, 16, pp. 111–21

Yearley, S. (1996) *Sociology, Environmentalism, Globalization: Reinventing the Globe*, Sage, London

Zalucki, M. (1997) Associate Professor, Department of Entomology, University of Queensland, pers. comm., July

Zechauser, R. and Viscusi, W. (1996) 'The Risk Management Dilemma', *Annals of the American Academy of Political and Social Science*, 545, pp. 144–55

Index

Also available from Scribe Publications

If you found *Altered Genes II* useful, you may also find the following recently published titles to be of interest.

You can also keep up with our new and forthcoming titles by viewing our website at www.scribepub.com.au

Important books
available at all good bookshops

REDESIGNING LIFE?
the worldwide challenge to genetic engineering

Editor:	Brian Tokar
ISBN:	0 908011 60 1
Format:	234 x 156mm paperback
Extent:	448pp
Category:	Environment/Technology
	Public Affairs

'This book provides a welcome perspective that is missing from all the biotechnology hype. For anyone concerned about the real implications and potential hazards of gene manipulation, this is an excellent starting point.'
—DAVID SUZUKI

Genetic engineering, animal cloning and new reproductive technologies are being promoted as the keys to a more productive agriculture, medical miracles, and solutions to pressing environmental problems. But growing numbers of farmers, scientists, and concerned citizens disagree, citing new hazards to our health and the environment, along with troubling ethical questions.

Redesigning Life? offers a comprehensive, global examination of the hidden hazards of the new genetic technologies and the emergence of worldwide resistance. Twenty-six internationally respected critics offer their analysis of the issues, their social and ethical implications, and the stories that lie behind the headlines that have brought genetic engineering to the forefront of public controversy.

The editor
Brian Tokar has been an activist since the 1970s in the peace, anti-nuclear, environmental and green politics movements, and is currently a faculty member at the Institute for Social Ecology and Goddard College in Vermont. He is the author of two books and numerous magazine articles. He graduated from MIT with degrees in biology and physics, and received a masters degree in biophysics from Harvard University.

The contributors
Contributors include Beth Burrows, Mitchel Cohen, Martha Crouch, Marcy Darnovsky, Michael Dorsey, Alix Fano, Jennifer Ferrara, Chaia Heller, David King, Sonja Schmitz, Thomas Schweiger, Robyn Seydel, Hope Shand, Vandana Shiva, Richard Steinbrecher, Jim Thomas, and Brian Tokar.

GENETIC ENGINEERING, FOOD, AND OUR ENVIRONMENT

a brief guide

Author: Luke Anderson
ISBN: 0 908011 43 1
Format: 198 x 128mm paperback
Extent: 192pp
Category: Environment/Politics

'A fantastic resource. The very essence of the case against genetic engineering — clearly, simply, and authoritatively explained.'
—BOB PHELPS, DIRECTOR, AUSTRALIAN GENEETHICS NETWORK

If current trends continue, within ten years most of the foods we eat could be genetically engineered. While multinational corporations want us to believe that this food is safe, nutritious, and thoroughly tested, we are entitled to be sceptical. As some dissenting scientists have pointed out, our current understanding of genetics is extremely limited, and the technology of genetic modification of food carries inherent risks both for human health and the environment. Despite official assurances, the introduction of genetically engineered organisms into complex ecosystems is a global experiment with unpredictable and irreversible consequences.

This compact, comprehensive book aims to clarify the key issues that concern people about genetic engineering. It answers questions such as:

* What is genetic engineering?
* Why are genetically engineered foods being introduced?
* What are the implications for health, farming, and the environment?
* Is genetic engineering needed to feed the growing world population?
* Why are living organisms being patented?
* Who is making the crucial decisions about the future of our food supply?
* What can you do if you are concerned about these issues?

The author
Luke Anderson is an English journalist, speaker, and campaigner who specialises in issues related to genetic engineering. He is a consultant to the UK Soil Association's genetic engineering campaign, and has written on the subject for other environmental organisations such as Greenpeace International.

He gave evidence to New Zealand's royal commission into genetic engineering in 2001.